POWER PLANT PERFORMANCE

NUCLEAR AND COAL CAPACITY FACTORS AND ECONOMICS

POWER PLANT PERFORMANCE

NUCLEAR AND COAL CAPACITY FACTORS AND ECONOMICS

CHARLES KOMANOFF

FOREWORD BY CHARLES WARREN

COUNCIL ON ECONOMIC PRIORITIES

NEW YORK AND SAN FRANCISCO

Research Assistants: Robert Urquhart and Jay Kita

Statistical/Computer Programming: Philip Levine and Walter Bourne

Editor: Fred Armentrout

Production and Design: Maris Cakars

Charles Komanoff, Energy Projects Director of CEP, has worked on energy policy issues since 1971. He co-authored *The Price of Power*, CEP's study of environmental pollution in the electric utilities industry, published by MIT Press. He was an economic analyst for the New York City Environmental Protection Administration, and has served as consultant and expert witness for numerous state agencies and environmental organizations. He is a member of the Federal Energy Administration's Energy Finance Advisory Committee. He is an honors graduate of Harvard College in Applied Mathematics.

International Standard Book Number: 0-87871-004-3
Library of Congress Catalog Card Number: 76-50521

Council on Economic Priorities,
84 Fifth Ave., New York, NY 10011

Printed in the United States of America.
Printed on 70% recycled paper.

Acknowledgments

This study was made possible by the contributions of many dedicated individuals.

Alice Tepper Marlin and Stephen K. Moody conceived the study, raised the necessary funds, and consistently provided valuable editorial and substantive review.

Research Assistants Robert Urquhart (1975) and Jay Kita (1976) diligently did most of the data gathering and computation. Philip Levine, principal statistical consultant in the early stages of the study, set up the regression models and computer programs. Walter Bourne skillfully handled computer programming in the later stages.

Fred Armentrout performed major stylistic editing of the manuscript. Maris Cakars handled copy-editing, design and production. Laura Boyer and Caroline Helmuth skillfully and patiently typed the manuscript through numerous revisions. Wendy Schwartz and Marcia Hornstein assisted greatly in the production.

Ernst Habicht of the Environmental Defense Fund and attorney Ken Semmel contributed ideas and criticism which significantly shaped the overall scope of the project. We are most grateful for their input.

John Berger of Friends of the Earth formulated the definition of capacity performance used for coal-fired plants. Maria Schafranek of Public Service Electric & Gas helped devise formulas to measure capacity performance. Gene Cramer of Southern California Edison and Dale Bridenbaugh of MHB Associates provided valuable criticism of Chapters 3-5.

A partial listing of others who contributed ideas and suggestions follows: David Comey, Citizens for a Better Environment; Alexander Gakner, FPC; Raymond E. Gerson, consultant; Jim Harding, California Energy Resources Commission; David Huettner, University of Oklahoma; Evan Kovacic, FEA; Robert Nelson, FEA; Kerry Smith, SUNY-Binghamton; Peter Stern, Northeast Utilities; Sally Streiter, NERA; Bernie Tennenbaum, FPC; Lee Thomas,

Vermont-American Corp. We also appreciate the efforts of the 60 electric utilities which provided us with detailed capacity factor information not available in FPC or NRC records.

None of the foregoing individuals are responsible for the accuracy of the data or findings in the study.

This is the first CEP study funded entirely by directed grants and consulting fees. Major grants were provided by the New York Community Trust, The LARAS Fund, the Shalan Foundation, Marion and Warren Weber, Lucy B. Lemann, the Conservation and Research Foundation, Inc., and several CEP members. We deeply appreciate their support.

During the research and writing of this study, CEP served as consultant on nuclear and coal capacity and economic issues to the Riverhead First Committee (Long Island), the Georgia Power Project, the Connecticut Public Utilities Control Authority, the New Jersey Department of Public Advocate, and the New Mexico Attorney General. The Federal Energy Administration has purchased the capacity factor data base developed in the study.

Charles Komanoff
November, 1976

TABLE OF CONTENTS

Appendices

Foreword

Contradictory data and conflicting claims frequently mark the debate over the economics of nuclear power. Plant operating reliability—a key issue in the debate—is satisfactory and improving, say nuclear proponents. It is the Achilles heel of nuclear economics, say the critics.

It has fallen to Charles Komanoff and the Council on Economic Priorities to unravel the tangled mass of claim and counterclaim, to separate hyperbole from fact, and to assemble all the relevant evidence.

In 1974 activists challenging nuclear power first suggested that the presumed economic advantage of nuclear power over coal was eliminated by the poor performance record of atomic power plants. Nuclear power was a good buy, they said, only if the plants were up and running most of the time. With a higher proportion of fixed costs than any other method of producing electricity, an idle nuclear plant soon becomes a heavy burden on the utility's financial ledger.

Activists saw in this argument a new chink in the armor of the nuclear industry. If the economic benefits of nuclear power were illusory, the chief reason for building atomic plants would vanish. The activists believed they had an argument which would strike a chord with utility planners, customers, and stockholders.

The industry response was to point out several defects in their critics' data and approach. They recognized quite clearly how damaging the capacity factor argument might be. The question was obviously not as simple as it had been portrayed. In the ensuing debate over the reliability of nuclear plants and alternatives to them, the issues and facts became even more clouded.

This CEP book puts the debate on solid ground. It develops the first complete data base of nuclear and coal plant operating reliability. It presents the first statistical equations that tie performance to plant design, size and age. And it describes the broad technical and regulatory factors that affect the reliability, and therefore the economics, of nuclear and coal generation.

Neither side in the original nuclear cost debate will be wholly pleased with the conclusions of CEP's study but both will benefit from it. Many small nuclear power plants have been performing well, perhaps to the disappointment of some critics. On the other hand, current utility plans emphasize the purchase of the size plants which CEP found to be performing most poorly, thus losing their cost advantage over coal. Moreover, CEP research demonstrates that the industry's "learning curve" argument, which asserts that nuclear plant performance improves over time so that future units will be more reliable than those currently operating, is largely not borne out by experience. The study thus calls into question a major justification for increased use of nuclear power.

But coal-fired power plants are no panacea either and are subject to their own reliability problems in large sizes. In their own economic self-interest, utility executives should pay close attention to these conclusions. While the book will not settle the debate, it has made a significant contribution by establishing which points are at issue and which are not.

The significance of this book goes beyond its subject matter. It demonstrates the value of an entirely new kind of organization to public policy debates—the non-profit public interest research group. While the power plant performance debate has been raging for two years, it took a research group independent of any traditional institution to do the type of careful and objective analysis required to resolve the points at issue. In the conventional view, problems of this nature would be resolved in the interchange between industry and its regulators. In this case, neither of the traditional parties were inclined to do the type of assessment which was needed. Fortunately, the Council on Economic Priorities provided the backstop. The major contribution made by Komanoff and the Council in producing this study will, hopefully, stimulate others to undertake similar service in support of better public decisions and debate. It is a trend which deserves encouragement, as this book demonstrates.

Charles Warren
46th Assembly District
Chairman of the Assembly Committee
on Resources, Land Use and Energy
California Legislature

1: Summary of Findings

Statistical Findings on Nuclear Capacity Factors (Chapter 3)

a. *Average:* Commercial U.S. nuclear plants have operated at a 59.3% average capacity factor through 1975, far below industry and government projections of 70%-80%. The sample includes all reactors of 450 Megawatt (Mw) capacity installed between 1968-1974, operating from 1968-1975.

b. *Size:* Nuclear capacity factors have declined with increasing unit capacity. Pressurized Water Reactor (PWR) capacity factors decline 3.4 percentage points per 100 Mw increase in capacity; Boiling Water Reactor (BWR) capacity factors decline 3.3 percentage points per 100 Mw.* Almost all nuclear units now planned are over 1000 Mw, compared to the 681-Mw average capacity to date.

c. *Maturity:* PWR's "mature," in that their capacity factors increase with age. BWR's, however, have shown no similar tendency.

d. *Type:* PWR's have operated at an average capacity factor of 61.6%—almost six percentage points higher than the 55.9% BWR capacity factor average.

e. *Learning:* Nuclear technology has not developed a capacity factor "learning curve," in that recently installed units have not operated at higher capacity factors than comparable earlier units at similar stages of maturity.

f. *Duplicates:* "Duplicate" PWR's—installed about a year after identical units at the same site—have operated at higher capacity factors than their predecessors. But small sample size reduces the significance of this finding. Data on duplicate BWR's was too limited to read for this finding.

g. *Range:* Lifetime capacity factors of nuclear units vary widely. The unit high was 77.2%, the low was 14.0%. Reactors operational at least two full calendar years have lifetime capacity factors ranging from Point Beach Unit #2's 75.9% to Palisades' 23.2%.

*This finding is not attributable to the lesser average age of large nuclear units, as compared to small units, since the statistical equations accounted for the effect of unit age along with that of unit size.

TABLE 1.1: 10-YEAR PROJECTED NUCLEAR CAPACITY FACTORS

Unit Size	PWR	BWR
600 Mw	66.2%	59.6%
800 Mw	59.4%	53.0%
1000 Mw	52.7%	46.4%
1150 Mw	47.6%	41.5%

h. *Utilities:* Wisconsin Electric Power, which operates Point Beach 1 and 2, has the highest utility average capacity factor: 71.6% over eight unit-years. Consumers Power, which operates Palisades, has the lowest: 23.2% over four unit-years.

Commonwealth Edison of Chicago, the nation's most experienced nuclear utility, has had 16 unit-years, or 15% of the total U.S. experience with commercial nuclear units. It has only a 53.3% total average capacity factor. This is six percentage points less than the U.S. nuclear average. However, Commonwealth's plants—large BWR's and very large PWR's—have capacity factors comparable to similar units of other utilities.

i. *Manufacturer:* Reactors manufactured by Westinghouse cover four-fifths of the PWR reactor-years in the CEP data base. They operate at an average capacity factor of 64.7%. This is almost nine percentage points higher than the 55.9% average for BWR's, all of which are manufactured by General Electric.* Capacity factors of PWR units manufactured by Combustion Engineering and Babcock & Wilcox have been less than those of Westinghouse units, but their production experience is too limited to draw meaningful capacity factor comparisons.

Table 1.1 projects average nuclear capacity factors for the first ten years of operation from capacity factor "trend equations" based on operating data. But caution should be exercised in using the table to project nuclear capacity factors, because of the following limits to accuracy:

The sample is small: 38 units, 110 unit-years of operation. Larger units (800 Mw or more) account for only one-third (36) of the unit years.

The sample is young: averages 2.9 years commercial operation per unit by the end of 1975. Only two units have operated eight years, and only two 800+ Mw units have operated four years.

Much variation in capacity factor data is not explained by unit size and age variables used in statistical analysis.

Changes in regulation, economics, and technology since units in the CEP data base were designed and constructed may affect future capacity factors in ways not now apparent.

However, Table 1.1 is biased toward high nuclear capacity projections because:

It is based upon nuclear plant operation only through the end of 1975.

*Roughly three percentage points of the 9% difference between Westinghouse and GE capacity factors are statistically attributable to the larger average size of GE units, based on the inverse relationship between nuclear capacity factor and unit size.

Inclusion of data on all 48 commercial nuclear units in the first half of 1976 (including 10 units which entered service in 1975) reduces U.S. cumulative average capacity factor from 59.3% to 58.1%. Such inclusion, seasonally adjusted, also reduces projected factors for unit sizes 700 Mw and above, reducing 1000-Mw PWR's by two percentage points and BWR's by 6%.

It omits "break-in" months of commercial operation prior to the first New Year's Day in which the unit was in service.

It uses the Nuclear Regulatory Commission's definition of design capacity, rather than the Federal Power Commission's definition of maximum nameplate generator capacity. The FPC definition has been used for coal-fired units, and would have decreased the average nuclear capacity factor by one percentage point (see Appendix B for capacity factor definitions).

It extrapolates current capacity factor trends and draws them out to the tenth year of operation, but assumes no reductions connected to plant aging. Operating reliability of the three pre-commercial 200-Mw nuclear plants has deteriorated after their tenth year.

Factors Affecting Nuclear Plant Capacity Factor (Chapter 4)*

a. The average 40.7% "lost" capacity factor at a typical nuclear plant, at the 59.3% performance average, is accounted for as follows:

1. *Scheduled Outages:* 12-18 month refueling and maintenance cycles and occasional other maintenance account for 18% average lost capacity factor. This is two to three times the loss anticipated in industry projections.

2. *Equipment Malfunctions:* equipment failures that cause power reduction or complete shutdown account for 17½% lost capacity factor. 10½% results from plant shutdowns; 7% from partial power reductions.

3. *Regulatory Restrictions:* NRC-mandated inspections or power deratings to comply with license specifications result in 4% lost capacity factor.

4. *Load-following and Time to Come to Full Power:* deliberate power reductions due to low system electric demand, and limits on rate-of-power acceleration to full capacity, together account for 1% lost capacity factor.

b. PWR's have been troubled by equipment failures in turbines and steam generators. BWR's have been plagued by failures of nuclear fuel and by extensions of refueling and maintenance outages to repair reactor core equipment.

c. No single cause dominates nuclear capacity problems. Fuel failures, the largest single category in equipment malfunction loss, account for only 3.5% lost capacity factor. 5.2% is so varied as to defy coherence to a single category. Similarly, no one cause explains refueling and maintenance loss.

Future Performance of Nuclear Units (Chapter 5)

a. *Size:* The lower average operating reliability of large nuclear units compared with small units results from a combination of (i) more components of

*Findings in this section are based on a study of nuclear outages by M.E. Lapides and E.L. Zebroski of the Electric Power Research Institute (see Chapter 4).

fixed size; (ii) more sets of components of fixed size in more demanding service conditions; (iii) more large components inherently more difficult to assemble and repair.

b. *Maturity and Age:* The strong tendency of commercial PWR's to operate at increasing capacity factors with increasing unit age is based only on operation through age 8. The sharply reduced capacity factors of pre-commercial 200-Mw reactors after age 10 are due to wear rather than regulatory restrictions. However, this performance drop cannot be applied with confidence to predictions for current reactors.

c. *Learning:* Nuclear plants may improve capacity factors as operating experience results are fed back into the design, construction, and operation of new plants. Increased economic incentives should result in capacity factor increases via higher construction and maintenance expenditures. Since unit duplication has increased PWR capacity factors, reactor design standardization, absent an unforeseen generic defect, may lead to higher capacity factors for new reactors.

d. *Impediments to Improvement:* Potential increases in nuclear capacity factors may be limited by reactors' technical complexity, missing clues to causes of outages, and possible increases in regulations on construction and operation. In-plant occupational exposure to radioactivity, a significant cause of prolonged outages, may result in further capacity factor reductions, unless minimized by more shielding or more effective design.

Statistical Findings on Coal Capacity Performance (Chapter 6)

a. *Average Capacity Factor:* Commercial U.S. coal-fired power plants have operated at an average capacity factor of 66.9%. The sample includes all 250 coal-fired units of 100+ Mw installed between 1960-1972 and operating from 1961-1973 (1641 unit-years).

b. *Average Capacity Performance:* Unlike nuclear units, many coal-fired units are not fully "base-loaded"—operated at constant full available capacity. "Capacity performance," a measure of operating reliability which corrects capacity factor data for "load-following," approximates the capacity factor at which a unit would have operated had it been fully base-loaded. Average capacity performance of the 250 coal-fired units is 74.8%.

c. *Type:* "Supercritical" coal units, in which the boiler is maintained at high pressures to increase thermal efficiency, have operated at about 4% higher capacity performance levels than "subcritical" units above 400 Mw. Higher performance of supercritical units is more pronounced above 600 Mw.

d. *Size:* Coal-fired capacity performance is inversely related to unit size. It declines at an average rate of 1.1 percentage points per 100 Mw capacity increase for supercritical units; at 5.6 percentage points per 100 Mw capacity increase for 100-399 Mw subcritical units; and at 2.2 percentage points per 100 Mw increase for 400+ Mw subcritical units.

e. *Maturity:* Coal-fired units take a full calendar year of commercial operation to "mature" to highest capacity factor, or 1-2 years for 600+ Mw units. Capacity factors of subcritical units start a decline at the 5th to 7th year of operation. Supercritical units and 600+ Mw units of both types, show in-

TABLE 1.2: 10-YEAR PROJECTED COAL UNIT CAPACITY FACTORS

Unit Size	Subcritical Units FGD†	Subcritical Units No FGD	Supercritical Units FGD	Supercritical Units No FGD
200 Mw	69%	73%	—	—
400 Mw	57-61%	60-65%	59-67%	62-71%
600 Mw	59-62%	62-65%	63-66%	66-70%
800 Mw	54-56%	57-59%	60-63%	63-66%
1000 Mw	49-50%	52-53%	57-63%	60-66%

†FGD refers to flue gas desulfurization systems ("scrubbers").

creased capacity factors through age 6. Data beyond that point is limited.

f. *Coal Grade:* Capacity performance is inversely related to coal sulfur content. It declines at a 1.8% rate for each 1% sulfur content rise. However, it is positively related to coal ash content for units under 600 Mw, at a 1.4% rate for each 1% increase. Capacity performance is generally not related to coal heat value or moisture content.

g. *Vintage:* Performance of subcritical coal-fired units under 600 Mw declines for those installed more recently in the 1960-1972 sample period. This is not observed for 600+ Mw units or supercritical units, as there were so few built prior to the late 1960's. Variation in capacity performance results from different measuring standards as well as actual performance.

h. *Duplicates:* "Duplicate" supercritical coal units of 600+ Mw capacity operate at higher capacity factors than their predecessors. Other subsets of the coal sample were not tested due to limited duplicate units or difficulties in identifying them as such.

i. *Utilities:* American Electric Power's supercritical 600+ Mw coal units have had capacity performance averages 9¼% higher than those of comparable units in the industry. Conversely, Pennsylvania Electric Company's units average 10½% less than comparable ones. Tennessee Valley Authority's subcritical 600+ Mw units have ranked 9½% more, and Commonwealth Edison's 10% less, than units of comparable size, age, and coal quality.*

Table 1.2 projects average coal capacity factors for the first ten years of operation from capacity performance trend equations based on operating data.

However caution should be exercised in using this table to project coal capacity factors because of the following limits to accuracy:

It applies to fully base-loaded plants. Capacity factors would be less for plants which "load-follow."

It assumes 2% sulfur content. Capacity factors would be less for plants burning higher-sulfur coal, greater for lower-sulfur coal.

Capacity performance data was computed from capacity factors by applying average rather than plant-specific estimates of load-following. This tends to

*Coal capacity averages are not otherwise given by utility since company variations in load-following may distort performance figures.

mask capacity performance differences related to variables other than unit size—such as age, fuel quality, year of construction, and utility owner's operating patterns.

Changes in regulation, economics, and technology since units in the CEP data base were designed and constructed may affect future capacity performance in ways not now apparent.

Factors Affecting Coal-Fired Plant Capacity Performance (Chapter 7)

a. *Design Margins:* Lower capacity performance of recent sub-600 Mw subcritical coal units is partially due to reduced capability margins with resultant failure to meet design capacity requirements.

b. *Coal Grade:* Sulfur content capacity performance loss results from fouling by sulfur-compounds formed in coal combustion. Capacity performance is also inversely related to low threshold coal ash-softening temperature, but loss can be minimized by proper match of boiler design to coal grade.

c. *Size:* Lower capacity performance of large coal units results from more components, more demanding service conditions, or larger components more difficult to assemble and repair. The milder effect of unit size on coal than on nuclear performance is partially due to less stringent regulation which permits operation of coal plants despite equipment breakdown.

d. *Type:* Supercritical boilers have operated at higher levels than subcritical, despite more complex design.

e. *Age:* Average decline in coal capacity performance in Ages 10-13 is only 1% per year, compared to decline several times greater for three prototype nuclear plants after Age 10.

f. *Utility:* Superior performance of AEP large units results from conservative design practices and a stepped construction program that builds on successes of smaller units. Pennsylvania Electric's poor large unit capacity performance results from inexperience in that size range and inadequate boiler design margins.

Future Performance of Coal-Fired Plants (Chapter 8)

a. *Incentives:* There is more economic incentive for improving coal plant capacity factors than existed when coal units in the CEP data base were installed. Still, the incentive increase is less than for nuclear plants. This incentive will likely be translated into more conservative design and construction practices to improve reliability.

b. *Duplication:* Unit duplication and, to a lesser extent, design standardization can improve coal-fired capacity performance, though potential is probably less than that for improvement of nuclear plants.

c. *Scrubbers:* Scrubbers will drop capacity performance of new coal-fired units below trend equation projections by about 5%, or 3-3½ percentage points for large units.

d. *Regulatory Impact:* Since 1970, environmental and mine safety regulations have indirectly led to reduced capacity performance in many coal-fired plants. Company compliance efforts have changed moisture and ash characteristics of coal burned in boilers and have shut down plants to upgrade pollution control equipment.

e. *Improvements:* Anticipation of regulations prior to design and construction of future plants should reduce outages related to air quality and mine safety regulation. Utility action toward improved matching of boiler specifications to coal grade burned and closer adherence to coal quality specifications by suppliers should reduce outages in boilers and related steam-generating equipment.

Optimal Power Plant Size (Chapter 9)

a. For both nuclear and coal-fired plants capital costs decrease for large units, but so do capacity factors. Based on capacity factor trend equations, unit sizes at which capital charges are lowest per kwh generated are: 600 Mw for PWR and BWR nuclear units, and 650 Mw for subcritical and supercritical coal units.

b. Optimum size of new power plants is smaller than that of the lowest capital charge per-kwh units. Smaller units enhance unit duplication or siting flexibility, reduce capacity reserve margins, and cut costs of fuel, operating and maintenance. Optimal unit size ranges are 350-700 Mw for nuclear units and 350-800 Mw for coal.

c. The average capacity of nuclear units to be installed between 1975-1984 is 1064 Mw. The typical nuclear unit being planned is 1150 Mw. Based on the capacity factor trend equations, the per-kwh capital charges of 1150-Mw nuclear units will exceed those of 600-Mw nuclear units by 17% for BWR's and 13% for PWR's, despite capital economies of scale in large units. If the inverse relation between nuclear unit size and capacity factor persists, the very large nuclear units will be considerably less economical than those (450-800 Mw) which now comprise most U.S. nuclear plant experience.

d. The average capacity of new fossil units being installed by U.S. utilities from 1975-1984 is 457 Mw. The typical new fossil unit planned is in the 500-600 Mw range. Most coal-fired units being added by U.S. utilities are within the size range in which the optimal combination of low capital cost and high operating reliability is achieved.

Economic Evaluation of Nuclear and Coal-Fired Units (Chapter 10)

a. Based on standard utility assumptions—75% nuclear and coal capacity factors, 71% for coal with scrubbers—new 1150-Mw nuclear units have the lowest total generating costs for an Eastern U.S. site, with 600-Mw coal units 23%-37% more expensive. However, based on different capacity factor projections from the CEP trend equations, but using standard cost assumptions for other parameters, new 1150-Mw PWR's and BWR's will be 16% and 31% more expensive, respectively, than 600-Mw supercritical coal units using Eastern low-sulfur coal, without scrubbers. If scrubbers rather than low-sulfur coal are used, 1150-Mw BWR's will be 12% more expensive. PWR and Eastern coal unit costs will be equivalent.

TABLE 1.3: 1984 NUCLEAR AND COAL GENERATING COSTS†

	Generating Cost (Mills/Kwh)	Ratio to 1150-Mw PWR
Eastern Coal		
Low-Sulfur		
Mine-Mouth	41.3	0.89
Transported	44.8	0.97
High-Sulfur		
Mine-Mouth	48.6	1.05
Transported	51.9	1.12
Western Coal		
Mine-Mouth	41.8	0.85
Transported	45.5	0.93

†Costs assume capacity factors from CEP trend equations, with 7½ percentage points added for nuclear. Eastern coal case requires low-sulfur coal or scrubbers, but not both. PWR cost for Eastern case is 46.2 mills; BWR cost is 50.9 mills. Western coal is low-sulfur but still requires scrubbers. Western PWR cost is 49.0 mills; BWR cost is 54.0 mills (see Chapter 10 for all cost assumptions).

b. A nuclear improvement 7½ percentage points over the CEP capacity factor trend equations is required to bring the generating costs of 1150-Mw nuclear units down to the level of 600-Mw supercritical Eastern coal units. With this assumed nuclear improvement but no coal capacity factor improvement, low-sulfur Eastern coal without scrubbers is still cheapest. PWR's add 3%; BWR's add 14%; high-sulfur coal with scrubbers adds 17%.

c. For a Western site, 1150-Mw PWR's are 8% and BWR's are 19% more expensive than 600-Mw supercritical coal units. This assumes: low-sulfur coal units with scrubbers; use of low-sulfur coal; all units with dry cooling towers; and a 7½% capacity factor improvement for nuclear units.

d. Cost comparisons assume coal must be transported from mine to plant. Relative coal-to-nuclear cost is reduced by about 10% for mine-mouth plants.

e. "Early installation" of very large nuclear plants—construction prior to their need for adequate system reliability—will create no overall cost savings as long as capacity factors remain under 55%.

Table 1.3 summarizes relative costs of 600-Mw supercritical coal units and 1150-Mw PWR's. These are the most likely designs and sizes for new coal and nuclear plants.

Major Policy Implications of the Study Findings

a. The study projections of capacity factors below 70% for conventional coal-fired plants and below 60% for nuclear plants improve the economic feasibility of alternatives such as energy conservation, solar energy, and advanced power technologies.

b. The study projections of nuclear capacity factors significantly below those of coal, based on currently planned unit sizes, eliminate nuclear power's national cost advantage over coal. Coal is competitive with nuclear for new plants in the Northeast, and more economical elsewhere, even with the assumed 7½% nuclear capacity factor improvement.

c. A return to smaller unit sizes (400-800 Mw) could improve nuclear power's competitiveness with coal, based on negative correlation between capacity factor and unit size.

d. High-sulfur coal cleaning gains feasibility as an alternative to scrubbers, based on negative correlation between sulfur content and capacity factor, and reliability impact of scrubbers.

e. Power plant standardization appears likely to improve capacity factors, especially for nuclear plants, without increasing capital and operating costs.

f. Overall, postponing commitment to new generating facilities, where this is permitted by reduced load growth, may reduce generating costs. Postponement would also facilitate choice of more reliable plants through further data base accumulation, capacity factor trend identification, and engineering improvement.

2: Introduction—The Debate Over Power Plant Capacity Factors

This study addresses the economics of future electric supply in the United States, as affected by the capacity factor* or performance reliability of nuclear and coal-fired power plants. It concludes that new large nuclear plants are more likely to operate at 45%-60% capacity factor levels than in the 70%-80% range projected by the electric utilities and reactor manufacturers who comprise the "nuclear industry." These latter projections rely on optimism unsupported by the actual operating experience of U.S. nuclear plants.

The assumption of high capacity factors is critical to arguments for the major nuclear power expansion sought by utilities and supported by federal energy policy makers. New nuclear construction is predicated on nuclear power's alleged economic advantage over: new coal-fired plants, additional generation by existing fossil-fuel plants, or energy conservation measures. However, this advantage is extremely dependent upon attainment of high nuclear plant capacity factors. At low capacity, nuclear fuel cost savings are insufficient to recover high nuclear plant construction costs. On a per-kilowatt hour basis, nuclear generating costs at 65% capacity factor are 18% higher than at 80% capacity factor. At 50% capacity factor, they are 46% higher.

Commercial nuclear plants operating in the U.S. range from 450 to 1100 Megawatt (Mw) capacity. They have operated from 1968 through 1975 at only 59% average capacity factor. Plants under 800 Mw averaged 64%; larger plants averaged 51%. Most large nuclear plants have not been in service as long as smaller units and so have been affected more by break-in problems. However, statistical analysis demonstrates that, regardless of plant age, nuclear capacity factors decline with increasing plant size. Most nuclear plants under construction and almost all currently planned are greater than 1000 Mw capacity.

Despite this record, the nuclear industry almost uniformly projects 70%-80% capacity factors for new plants. These projections are the foundation of utility estimates of nuclear power cost advantages submitted in reactor

*Capacity factor is a power plant's actual electrical generation as a percentage of its maximum possible generation, over a period of time. It represents the average rate at which a plant is producing power, as a percentage of its maximum or "design" capacity.

licensing applications before the federal Nuclear Regulatory Commission and state regulatory authorities. They are the basis of assertions by reactor vendors and the Federal Energy Administration that each 1000-Mw nuclear plant will displace present use of 11 million barrels of oil per year. They were presented as authoritative by the federal Energy Research & Development Administration in its widely publicized estimates of the cost of the proposed curbs on nuclear power which were ultimately rejected by California voters in June 1976.

Despite its critical importance, capacity factor has been the missing element in the national debate over nuclear economics. Controversy has focused largely on capital costs, fuel costs and "fuel cycle" issues such as uranium availability and plutonium recycle. Recently, however, low industry capacity averages and erratic performance of many individual plants have generated skepticism about utility capacity factor projections.*

The industry's public response to criticsm of its low capacity factor record has been four-fold. First, it has sought to focus attention selectively on the best performing plants—four of 38 nuclear units entering commercial service before 1975 have better than 70% lifetime capacity factors through that year; ten others have reached 70% for at least one calendar year.

Second, some industry spokespersons have inflated statistics by applying less stringent performance standards than capacity factor based on plant design capacity. "Availability factor" measures only the percent of time a plant is operating at any capacity, and thus ignores partial power reductions common to nuclear plants. "Maximum dependable capability" yields misleadingly high capacity factors by reducing capacity targets when plant capability is limited by mechanical or regulatory factors which have been generic to nuclear technology.**

Third, utilities and reactor vendors project significant capacity factor increases linked to maturation of individual plants and to improvements in reactor technology as a whole. Yet the industry has never offered estimates of the extent of "maturation" or the "learning curve" allegedly followed by nuclear plants to date. In actuality, only one of two reactor types improves with age, and neither shows statistically significant increased capacity factors linked to later installation date, if unit size and age are kept constant. Moreover, nuclear capacity factors decrease markedly as size increases. This suggests that future large reactor performance will even fall short of current units.

Fourth, the industry cites operating difficulties of coal plants—the only presently available large-scale generating alternative to nuclear plants—in arguing that nuclear plants are equally reliable and more economical overall. This study demonstrates that large coal plants (400 to 1150 Mw capacity) operated at only 58% capacity factor between 1968-1973. But their achievable

*Industry-wide nuclear capacity factor data was first critically analyzed by David Comey in September 1974 (see Bibliography).

**Typical of this approach is General Electric's claim that "BWR (Boiling Water Reactor) plants continue to perform well (in 1975, the best achieved 91% availability; the top six performers averaged 87% availability)" in *Electrical World*, September 15, 1976 (p. 79). Actually, the 14 BWR's in commercial service throughout 1975 averaged only 51% capacity factor that year, far below any reasonable definition of good performance. No BWR has *ever* reached 80% capacity factor in any calendar year.

capacity factor averaged 63%, after accounting for intentional output reductions during low power demand periods. In addition, overall coal generating costs are only three-fifths as sensitive to low capacity factors as costs of nuclear plants, due to coal's lower capital costs which reduce impact of idle capacity. Finally, most coal units under construction and planned are only half the megawatt capacity of new nuclear plants. This suggests a capacity factor advantage for future coal plants over nuclear, even though coal plants will probably not meet utility capacity factor projections.

Utility projections of 70%-80% capacity factors for both coal and nuclear plants are based on consistent operation by most older, smaller fossil units and some modern coal and nuclear installations. Large modern power plants, however, are more complex than earlier designs. New plants are subject to stricter environmental and safety regulations. While some plants are consistently reliable, others are often shut down for maintenance and repair, especially very large capacity nuclear plants. Power engineers have yet to isolate many factors affecting performance, so that capacity factors of specific plants cannot be pinpointed ahead of time.

However, capacity factors can be statistically correlated with plant size, age, design, and installation date to project average capacity factors for groups of plants. The correlations in this study are the first power plant capacity factor trend equations developed, and are based on the first complete year-by-year capacity factor data base to be compiled for either nuclear or coal units. The correlations indicate average capacity factors of 45%-50% for nuclear plants and 60%-70% for coal plants of typical megawatt capacities now planned—1150-Mw for nuclear, 600-Mw for coal. This assumes today's technology. Design and operating advances may boost future averages, especially for nuclear plants. Even so, improvements will be limited by tight regulation, gaps in engineering knowledge, and reliability problems intrinsic to large plants and perhaps to aging nuclear plants.

An accurate assessment of nuclear and coal capacity reliability and cost is essential to U.S. energy planning. Nuclear and coal together account for slightly over half of current U.S. electric generation, and their share is increasing rapidly. Oil and gas are no longer available to fuel new power plants, and are beginning to be phased out of traditional uses such as boiler fuel and industrial process heat. Solar-derived sources appear to have considerable future promise—especially for low-quality heat applications—but are not now developed for wide commercial application. Energy conservation—increasing the efficiency of energy use—is slowed by regulatory complexity, inadequate consumer information, and marketplace disincentives.

The result is a projected resumption in growth in energy use after a two-year hiatus following the 1973-74 Arab oil embargo. An increasing share of U.S. energy is planned to be provided by electricity, with nuclear and coal accounting for virtually all new major ("base-load") capacity. Based on current industry planning, nuclear capacity will increase from 40,000 Mw in 1975 to 180,000 Mw in 1985 and 600,000 Mw in the year 2000. Coal capacity will increase from 180,000 Mw currently to 300,000 Mw in 1985 and 550,000 Mw in 2000.

Sparked by environmental and safety concerns, expansion of nuclear capacity is the subject of intense national debate. Nuclear plant safety is an unresolved issue, with continuing controversy over the integrity of reactor mechanisms designed to prevent release of radioactive materials under accident conditions. The nuclear "fuel cycle"—stages from uranium mining through spent fuel reprocessing and waste disposal—potentially impacts both the biosphere and geopolitical security.

Coal use, however, also creates significant safety and environmental impacts. Underground coal mining is extremely hazardous, and strip mining is potentially destructive to land, especially in the arid West. Coal combustion releases toxic sulfur compounds (accounting for half from all U.S. sources), particulate matter, and nitrogen oxides. Accordingly, the "nuclear vs. coal" debate is often characterized by concern over significant increases in either type of capacity, and creates uncertainty for the utility manager.

Power supply expansion is viewed by electric utilities and much of business, government, and organized labor as vital to economic prosperity. Inadequate base-load capacity could raise energy costs, worsen unemployment, and deepen American dependence on imported oil, which currently accounts for 42% of U.S. oil consumption and 19% of overall energy consumption. According to this view, economic benefits from additional coal and nuclear supply outweigh environmental or social costs.

But low capacity factors increase power costs, and therefore reduce power expansion benefits. Utilities must build additional plants to meet demand, and operate less efficient plants while newer nuclear or coal units are shut down or only partially operating. Low capacity operation also thwarts efforts to reduce utility use of scarce oil and natural gas—one of the premises of nuclear and coal capacity expansion.

Per-kilowatt-hour (kwh) costs projected for future plants have tripled since the early 1970's, after decades of stable or even declining costs for new power plants. Nuclear and coal construction costs have skyrocketed as a result of inflation, additional environmental and safety design considerations, productivity declines, lengthened construction times, and higher interest rates. Fuel costs have spiraled with rising production expenses and greater supplier bargaining leverage in the wake of the OPEC-induced oil price hike. Poor capacity factors have added significantly to generating costs. Per-kwh costs of new plants greatly exceed those of existing plants,* even omitting substantial federal support through research and tax subsidies which are not reflected in utility costs.

Cost is a major influence on utility selection of new power plants. High public concern over rising electric costs has forced utilities to defend cost projections in hearings before state regulatory authorities and in debate with activist groups. Comparative nuclear and coal costs, and relative costs of conservation alternatives to new capacity, have been prominent issues in power facility siting hearings in many states.

Sweeping generalization about relative nuclear and coal costs may obscure regional or design differences. Nuclear plants can be built with pressurized or

*This reverses the historic "benign cycle" in the electric utilities industry, in which long-run marginal costs were stable or declining, enabling supply increases to reduce average costs. Currently the benign cycle applies only to some nuclear power displacement of oil- and gas-fired units and there the cost savings are at best marginal due to low capacity factors.

boiling water reactors. Coal plants use subcritical or supercritical (high-pressure) boilers, and require low-sulfur coal or scrubbers or both. Plant sizes range from several hundred megawatts to over 1000 Mw. Geographical location affects coal grade and cost and plant construction wage rates.

Proposed nuclear construction is concentrated in three areas of the U.S.: the Northeast, which is far from coal fields; the Southeast, where low wage rates reduce nuclear construction costs; and some Midwestern states where local coal is high-sulfur and therefore unusable in new plants without expensive scrubbers. In these cases, utilities anticipate sufficient nuclear cost advantage over coal to outweigh nuclear drawbacks of greater licensing hurdles, construction financing requirements, and uncertainties in fuel supply and regulation. However, these cost projections assume a nuclear performance potential that is not reflected in actual experience and which is unlikely to be realized in the future.

This study was undertaken in order to help resolve the nuclear/coal capacity factor and economics debate by:

1. Providing a complete capacity factor data base for modern U.S. nuclear and coal-fired units (Appendix A).
2. Statistically analyzing the data base to correlate capacity factor to plant design, unit size and age, and other key variables for both nuclear and coal (Chapters 3 and 6).
3. Identifying factors which most affect capacity factor and assessing the potential for improvement (Chapters 4, 5, 7 and 8).
4. Determining optimal unit size and design type for nuclear and coal-fired units, based on size effect on capacity factor and capital cost to date (Chapter 9).
5. Estimating and comparing total nuclear and coal generating costs, based on capacity factors anticipated for unit sizes planned by U.S. utilities (Chapter 10).

The major implications of the study's findings are as follows:

1. CEP projections of nuclear capacity factors significantly below coal, based on unit sizes currently planned, eliminate nuclear power's assumed national average cost advantage over coal. Coal is competitive with nuclear for new plants in the Northeast, and more economical elsewhere.
2. CEP projections of capacity factors below 70% for conventional coal-fired as well as nuclear plants improve the economic feasibility of alternatives such as energy conservation, solar space and water heating and advanced technologies such as fluidized bed combustion of coal and combined cycle plants.
3. A return to smaller unit sizes (400-800 Mw) could increase nuclear power's competitiveness with coal, based on the negative correlation between capacity factor and unit size.
4. Cleaning of high-sulfur coal gains economic feasibility as an alternative to scrubbers, based on negative correlation between sulfur content and capacity factor, and reliability impact of scrubbers.
5. Overall, postponing commitment to new generating facilities, where possible, may reduce ultimate generating costs. Moreover, postponement could facilitate more reliable plant design selection through further data base accumulation, capacity factor trend identification, and engineering improvement.

3: Statistical Analysis of Nuclear Capacity Factors

Section 3.1: Major Statistical Findings on Nuclear Capacity Factors

The operating record of commercial-size U.S. nuclear power plants is characterized by capacity factors lower than the industry-projected range of 70%-80%. There is also significant variation in plant capacity factors between units and between years. The 38 nuclear units in the CEP data base have operated for a total of 110 unit-years at an average capacity factor of 59.3% (with a standard deviation of 16.0%).*

Though the current average capacity factor is significantly below the 80% historically projected by most utilities, reactor vendors and nuclear plant architect-engineers (the major members and spokespersons of the "nuclear industry"), units in the CEP nuclear data base are not fully representative of future nuclear operation. The average nuclear unit-year in the data base is younger (2.9 years) than fully "mature" nuclear units, and smaller (681 Mw) than future nuclear units, which will be in the 1000-1300 Mw range. Many nuclear units have shown a tendency toward higher capacity factors with increased age. However, nuclear plants show a marked tendency toward lower capacity factors related to large unit size.

*59.3% is the average of the 110 unit-year capacity factors. It is the capacity factor of a typical reactor-year to date. Unit-year averages are used in this study unless otherwise noted. Size-weighted average capacity factor—total generation divided by total capacity—is 57.8%. This measures capacity factor of an average nuclear megawatt.

The CEP nuclear data base is in Appendix A, as are criteria used to define it. The data base includes 1968-1975 operation of all U.S. nuclear units 450 Mw and larger, which entered commercial service 1968-1974.

Nuclear capacity factor is measured by net generation and Design Electrical Rating reported to the NRC (see Appendix B). Average nuclear capacity factor is 58.4% (if capacity is measured as FPC Installed Generating Capacity) or 61.4% (if measured as Maximum Dependable Capacity reported to NRC).

The 59.3% mean capacity factor and 16% standard deviation suggest that 68% of 110 unit-years would fall within the capacity factor range, 43.3% to 75.3%. Actually, 71% of the nuclear capacity factors are within this range.

Statistical analysis of nuclear plant capacity factors yields the following conclusions:

1. Nuclear capacity factors have been inversely related to unit size (Mw capacity). Pressurized Water Reactor (PWR) capacity factors have declined at an average rate of 3.4 percentage points per 100 Mw increase in capacity; Boiling Water Reactor (BWR) capacity factors have declined at an average rate of 3.3 percentage points per 100 Mw.*

2. PWR's have demonstrated a tendency to "mature," in that the capacity factor of a typical PWR has generally increased as the unit has passed through early years of operation. BWR's, however, have shown no tendency to change capacity factor as units become older.**

3. Pressurized Water Reactors have operated at an average capacity factor of 61.6%—almost 6 percentage points higher than the average Boiling Water Reactor capacity factor of 55.9%.

4. Nuclear technology has generally not exhibited a capacity factor "learning curve," in that recently installed units have not operated at higher capacity factors than comparable earlier units at similar stages of maturity.

5. "Duplicate" PWR's—installed about a year after identical units at the same site—have operated at significantly higher capacity factors than their predecessors.*** Data on duplicate BWR's was too limited to read for this finding.

6. Lifetime capacity factors of nuclear units have varied widely. The unit high was 77.2%, the low was 14.0%. Reactors operational for two or more full calendar years have lifetime capacity factors that range from Point Beach Unit #2's 75.9% (Wisconsin Electric Power) to Palisades' 23.2% (Consumers Power).

7. Commonwealth Edison of Chicago, the nation's leading nuclear utility, has had 15% of the total U.S. experience with commercial nuclear units. Its nuclear plants have only a 53.3% total average capacity factor. This is six percent less than the U.S. nuclear average. However, Commonwealth's large BWR's and very large PWR's have, individually, operated at capacity factors similar to those of comparable units run by other utilities.

8. Reactors manufactured by Westinghouse provide four-fifths of the PWR reactor-years covered in the CEP data base. They operate at an average capacity factor of 64.7%. This is almost nine percent higher than the 55.9% average for BWR's, all of which are manufactured by General Electric.**** Capacity factors of PWR units manufactured by Combustion Engineering (CE) and Babcock & Wilcox (B&W) have generatlly been less than those of

*This finding is not attributable to the lesser average age of large nuclear units, as compared to small units, since the statistical equations accounted for the effect of unit age along with that of unit size.

**Conclusions about unit age and capacity factor are based upon units in the CEP data base, only two of which have accumulated as many as 8 calendar years of commercial operation.

***Small sample size reduces the significance of this finding.

****Roughly three percent of the 9% difference between Westinghouse and GE capacity factors are statistically attributable to the larger average size of GE units compared to Westinghouse, based on the inverse relationship between nuclear capacity factor and unit size noted.

Westinghouse units. But their production experience is too limited to draw meaningful capacity factor comparisons.*

Findings are quantified in capacity factor "trend equations" developed from the CEP data base (see Sections 3.3 and 3.4). The equations relate capacity factors to BWR unit size and PWR unit size and age. Projected from these equations, capacity factors for the first 10 years' commercial operation of hypothetical nuclear units are as shown in Table 3.1.

Interpretive Limits to Table 3.1

Caution should be exercised in using Table 3.1 or the trend equations to project nuclear capacity factors. The following are some of the limits to accuracy they impose:

a) Sample space is small: 38 units, 110 unit-years of operation. Larger units (800 Mw or more) account for only one-third (36) of the unit years.**

b) The sample is young. It averages 2.9 years commercial operation per unit by the end of 1975.*** Only two units have operated eight years, and only two 800+ Mw units have operated four years.

c) Trend equations satisfy minimum criteria for statistical significance but do not have full explanatory value, since most variation in unit-year capacity factor data is not explained by the variables chosen (unit size, age, and so forth). Some variation is due to erratic and lengthy outages (refueling and forced outages) in the calendar-year framework used to measure capacity factor in this study.****

d) Future nuclear plants will operate in a regulatory, economic, and technical information climate very different from those in the CEP data base. Regulation of plant operations is tightening. But more information on causes of plant outages promises to guide improved design and operating practices. Nuclear construction more often duplicates units at one site and incorporates greater design standardization. Duplicate PWR's may help increase nuclear capacity factors (see finding #5).*****

Moreover, a near tripling of the economic incentive to utilities for increasing nuclear capacity factors has followed on the wake of the Arab oil embargo, which sharpened the difference between nuclear and fossil per-kilowatt hour fuel costs.******

*The six B&W unit-years in the CEP data base have a high average capacity factor, while the eight CE unit-years have been low. However, data for the first half of 1976 has followed an opposite pattern.

**However, trend equation findings are confirmed by data from the first six months of 1976. This data adds 24 unit-years to the sample space, 13 of which are from units 800 Mw or larger (see Appendix D).

***The CEP data base excludes partial years of commercial operation prior to New Year's Day, so average age of sample units was just over three years at the end of 1975 (see Appendix A).

****Nuclear plants tend to be refueled on a 15 to 18-month schedule, so refueling shutdowns do not always coincide with a calendar-year. This produces annual capacity factor variations that cannot be correlated with unit size or age, though general levels may be so correlated.

*****Prospects for improved nuclear capacity factors are discussed in Chapter 5.

******The effective cost of nuclear plant outages is the resulting increase in the system fuel cost.

TABLE 3.1: PROJECTED AVERAGE CAPACITY FACTORS OF HYPOTHETICAL NUCLEAR UNITS

	PWR			BWR††
Unit Size	Age = 1	Age = 10	10-year Average†	
600 Mw	59.2%	69.2%	66.2%	59.6%
800 Mw	52.5%	62.5%	59.4%	53.0%
1000 Mw	45.8%	55.7%	52.7%	46.4%
1150 Mw	40.7%	50.7%	47.6%	41.5%

†The 10-year average capacity factors are "levelized" (computed on present-worth basis) rather than arithmetic averages (see Appendix F).

††BWR capacity factors are not projected as a function of their unit age because no statistically significant relationship exists.

High Capacity Factor Bias of Table 3.1

Table 3.1 is also biased toward high nuclear capacity projections *to date*. For instance:

a) It is based upon nuclear plant operation only through the end of 1975. Inclusion of data on 48 commercial nuclear units in the first half of 1976 (including 10 units which entered service in 1975) reduces U.S. cumulative average capacity factor (59.3% to 58.1%). Such inclusion would also reduce projected capacity factors for unit sizes 700 Mw and above, and would reduce capacity factors of 1000-Mw PWR's by two percentage points and BWR's by 6%.*

b) It omits months of commercial operation prior to the first New Year's Day in which the unit was in service. This "break in" period is characterized by low capacity factors.

c) It uses the Nuclear Regulatory Commission's definition of "design capacity," rather than the Federal Power Commission's definition of "maximum nameplate generator capacity." The FPC definition has been used in computing capacity factors of coal-fired units, but would have decreased the average nuclear capacity factor by one percentage point.

d) It extrapolates current capacity factor trends and draws them out to the tenth year of operation, but assumes no capacity factor reductions connected to plant aging. Operating reliability of power plants generally starts to deteriorate at or before the tenth year.**

*This reduction in projected capacity factors is based on trend equations in which data for the first half of 1976 is extrapolated to include an assumed improvement from the first to the second half of 1976 that reflects seasonal influences (described in Appendix D).

**Examples of this trend are the three nuclear plants in the 200-Mw class which have operated for at least 10 years and have each suffered significant capacity factor reductions beyond that point. These units are not in the CEP data base (see Chapter 5).

TABLE 3.2: LIFETIME CAPACITY FACTOR OF NUCLEAR UNITS

Unit	Utility	Capacity (in Mw)	Capacity Factor	#Years of Operation
1. Three Mile Island	Metropolitan Edison	819	77.2%	1
2. Point Beach 2	Wisc Elec Power	497	75.9%	3
3. Conn. Yankee	Northeast Utils	575	73.3%	8
4. Robinson 2	Carolina P & L	707	71.0%	4
5. Point Beach 1	Wisc Elec Power	497	69.0%	5
6. San Onofre 1	So Cal Edison	450	68.8%	8
7. Prairie Island 2	No States Power	530	68.4%	1
8. Kewaunee	Wisc Pub Serv	560	68.1%	1
9. Monticello†	No States Power	545	66.2%	4
10. Arkansas 1	Arkansas P & L	850	65.5%	1
11. Oconee 3	Duke Power	886	64.9%	1
12. Oyster Creek†	Jersey Central	650	64.5%	6
13. Oconee 2	Duke Power	886	64.0%	1
14. Turkey Point 4	Florida P & L	745	63.5%	2
15. Ginna	Rochester G & E	490	63.4%	5
16. Quad Cities 1†	Commonwealth Ed	809	59.9%	3
17. Oconee 1	Duke Power	886	59.8%	2
18. Vermont Yankee†	New England Elec	514	58.2%	3
19. Turkey Point 3	Florida P & L	745	57.7%	3
20. Quad Cities 2†	Commonwealth Ed	809	57.2%	3
21. Peach Bottom 3†	Philly Elec	1065	56.6%	1
22. Cooper†	Nebraska PPD	778	56.5%	1
23. Prairie Island 1	No States Power	530	55.2%	2
24. Nine Mile Pt 1†	Niagara Mohawk	610	55.2%	6
25. Maine Yankee	New England Elec	790	55.0%	3
26. Peach Bottom 2†	Philly Elec	1065	54.5%	1
27. Indian Point 2	Consolidated Ed	873	53.7%	2
28. Dresden 2†	Commonwealth Ed	809	53.4%	3
29. Millstone 1†	Northeast Utils	690	53.4%	5
30. Surry 2	Va Elec & Power	823	53.3%	2
31. Zion 2	Commonwealth Ed	1050	52.5%	1
32. Fort Calhoun	Omaha PPD	457	52.0%	1
33. Surry 1	Va Elec & Power	823	49.5%	3
34. Dresden 3†	Commonwealth Ed	809	49.4%	4
35. Pilgrim 1†	Boston Edison	670	49.0%	3
36. Zion 1	Commonwealth Ed	1050	45.6%	2
37. Palisades	Consumers Power	821	23.2%	4
38. Browns Ferry 1†	TVA	1098	14.0%	1

†Indicates BWR. All others are PWR.

TABLE 3.3: DISTRIBUTION OF LIFETIME NUCLEAR CAPACITY FACTORS

Lifetime Capacity Factor	Number of Units	
	All 38 Nuclear Units	26 Nuclear Units with 2 or more Calendar Years†
10-29.9%	2	1
45-49.9%	4	4
50-59.9%	17	12
60-69.9%	11	6
70-77.2%	4	3

†First-year capacity factors show high fluctuation due to break-in problems and omission of refueling and subsequent down-time in most cases.

TABLE 3.4: DISTRIBUTION OF NUCLEAR ONE-YEAR CAPACITY FACTORS

Capacity Factor	Number of Unit-Years
0- 9.9%	1
10-19.9%	1
20-29.9%	1
30-39.9%	11
40-49.9%	11
50-59.9%	23
60-69.9%	35
70-79.9%	20
80-86.4%†	7

†Highest single-year capacity factor (on a calendar-year basis) was 86.4% achieved by Conn. Yankee in 1974. A practical maximum capacity factor of roughly 92% exists for nuclear plants, based on a nominal outage of one month per year for maintenance and refueling. In practice, scheduled and forced outages have added to reactor down-time (see Chapter 4).

Section 3.2: Nuclear Capacity Factor Averages and Statistics

3.2a: Capacity Factor by Unit

Table 3.2 lists lifetime capacity factor (from first New Year's Day of commercial service) of the 38 nuclear units in the CEP data base, in decreasing capacity factor order.

Lifetime capacity factor ranges from 14.0% (Browns Ferry 1) to 77.2% (Three Mile Island), both with only one full year of commercial operation. Units aged two or more calendar years have a lifetime capacity factor range from 23.2% for 821-Mw Palisades (four years) to 75.9% for 497-Mw Point Beach 2 (three years).*

*If operation through June 1976 is included, the six nuclear units with the highest lifetime capacity factor (descending order) are: Point Beach 2, Connecticut Yankee, Robinson 2, Point Beach 1, San Onofre, and Monticello. See Table D2, Appendix D for complete list.

TABLE 3.5: AVERAGE LIFETIME NUCLEAR CAPACITY FACTOR BY OPERATING UTILITY

Utility	Number of Units	Number of Unit-Years	Average Unit-Year Capacity	Average Capacity Factor
1. Metropolitan Edison (Pa.)	1	1	819 Mw	77.2%
2. Wisconsin Elec. Power	2	8	497 Mw	71.6%
3. Carolina P & L (N.C.)	1	4	707 Mw	71.0%
4. So. California Edison	1	8	450 Mw	68.8%
5. Wisconsin Public Service	1	1	560 Mw	68.1%
6. Northeast Utilities (Conn.)†	2	13	619 Mw	65.7%
7. Arkansas P & L	1	1	850 Mw	65.5%
8. Jersey Central P & L	1	6	650 Mw	64.5%
9. Rochester Gas & Elec (N.Y.)	1	5	490 Mw	63.4%
10. Northern States Power (Minn.)	3	7	539 Mw	63.3%
11. Duke Power (N.C./S.C.)	3	4	886 Mw	62.1%
12. Florida P & L	2	5	745 Mw	60.0%
13. New England Elec.	2	6	652 Mw	56.6%
14. Nebraska Public Power Dist.	1	1	778 Mw	56.5%
15. Philadelphia Elec.	2	2	1065 Mw	55.5%
16. Niagara Mohawk	1	6	610 Mw	55.2%
17. Consolidated Edison (N.Y.)	1	2	873 Mw	53.7%
18. Commonwealth Edison (Chicago)	6	16	854 Mw	53.3%
19 Omaha Public Power Dist.	1	1	457 Mw	52.0%
20. Va. Electric & Power	2	5	823 Mw	51.1%
21. Boston Edison	1	3	670 Mw	49.0%
22. Consumers Power (Mich.)	1	4	821 Mw	23.2%
23. TVA	1	1	1098 Mw	14.0%

†Includes Connecticut Yankee and Millstone 1. The latter is owned and operated by Northeast Nuclear Co., which is owned principally by Northeast Utilities.

Excluding Browns Ferry 1 and Palisades, lifetime capacity factors of all remaining units fall within 45%-77%. Twenty-eight units are at 50%-69%. No unit has averaged 80%, despite projections of that figure by the AEC and industry in the past and some utilities even now.

Distribution of single-year capacity factor performance is also of some interest and is shown in Table 3.4 which demonstrates that nuclear one-year capacity factors follow a "log-normal" distribution* centered on the 60%-69.9% capacity factor range. The difference between the historic average

*In a log-normal distribution, the number of observations peaks at midpoint, has a sharp cutoff at a zero value, and tails off toward the high end of the distribution.

TABLE 3.6: AVERAGE CAPACITY FACTOR BY REACTOR MANUFACTURER

Reactor Manufacturer	# Units	# Unit-Years	Average Capacity†	Average Capacity Factor
Westinghouse	16	52	622 Mw	64.7%
Babcock & Wilcox (B&W)	5	6	869 Mw	65.2%
Combusion Engineering (CE)	3	8	764 Mw	38.8%
PWR Average	24	66	662 Mw	61.6%
General Electric	14	44	711 Mw	55.9%

†Average capacity is computed by weighting each unit's capacity by the number of years of operation in the CEP data base.

capacity factor of 59.3% and the median 60%-69.9% range with capacity factors below 40% omitted suggests averages could be improved if very poor years could be avoided.

Substantial variability in data between years and units (Tables 3.2 through 3.4) suggests it is inappropriate to use unit-years from selected reactors in selected years to support capacity factor projections. A few data points might be chosen to justify virtually any projection. It is necessary to use statistical methods applied to the entire data base (see Sections 3.3 and 3.4).

3.2b: Capacity Factor by Operating Utility

The 38 nuclear units are operated by 23 electric utilities. Fourteen utilities operate only one unit each. At the other extreme, Commonwealth Edison operates six units with a total of 16 unit-years. This accounts for one-seventh of the experience in the CEP nuclear data base.

Besides Metropolitan Edison with 77.2% nuclear capacity factor due to only one year of operation by its Three Mile Island unit,* Wisconsin Electric Power was highest. It had an average capacity factor of 71.6% for eight unit-years of operation (497-Mw Point Beach 1 and 2).

There is no significant relationship between capacity factor and utility experience with nuclear units, other than perhaps a slight tendency for utilities with greater experience to have higher capacity factors.** The exception is Commonwealth Edison. Its average capacity factor is 53.3%, versus 60.3% for all nuclear units minus Commonwealth's.***

*Three Mile Island was not refueled in 1975, but was shut down in early 1976 for an extended refueling and maintenance outage, dropping its lifetime capacity factor to 65.7% through June 30, 1976.

**This is statistically attributable to the tendency for individual PWR's to have higher capacity factors with increased age (see Section 3.3).

***Capacity factors of BWR's operated by Comm Ed were not statistically different from BWR's operated by other utilities. Thirteen of Comm Ed's 16 nuclear unit-years are accounted for by BWR's (see Section 3.4).

3.2c: Capacity Factor by Reactor Type and Manufacturer

A nuclear power plant is built around the *reactor,* or "nuclear steam supply system," which generates heat from nuclear fuel to produce steam. Steam then enters a turbine to turn a generator to produce electricity. Reactors are supplied to U.S. utilities by four different companies: Babcock & Wilcox (B&W), Combustion Engineering (CE), and Westinghouse (W) all manufacture Pressurized Water Reactors (PWR's), and General Electric (GE) which manufactures Boiling Water Reactors (BWR's).* The average capacity factors of nuclear plants grouped by the four reactor manufacturers are given in Table 3.6.

Westinghouse and GE reactors, which account for 87% of commercial U.S. nuclear operating experience through 1975, operate at significantly different average capacity factors. Westinghouse reactors operated at 64.7% average capacity factor, almost nine percent higher than 55.9% for GE reactors.**

The 65.2% average capacity factor of B&W reactors through 1975 and 38.8% average capacity factor of CE reactors do not accurately reflect current performance. This is due to small samples involved (six and eight unit-years, respectively) and other "quirks" in the B&W and CE data.

The B&W capacity factor probably overstates current cumulative capacity performance, since only two of six B&W unit-years in the data base included a refueling shutdown. Moreover, three B&W reactors that account for four of six unit-years, are owned by Duke Power (Oconee 1, 2 and 3), which does its own power plant construction and has a strong reputation for reliability.*** The high B&W reactor capacity factors largely reflect refueling schedules and engineering experience of Duke Power.****

*A fifth U.S. company, General Atomic—a joint subsidiary of Gulf Oil and Royal Dutch Shell—manufactures High-Temperature Gas-cooled Reactors (HTGR). HTGR design differs from PWR's or BWR's, in that it employs (i) *highly* "enriched" rather than *slightly* enriched uranium as fuel; (ii) helium gas rather than water as the heat exchange medium; and (iii) graphite rather than water as the neutron "moderator."

General Atomic is not included in this study since the only U.S. commercial-size HTGR—the 330-Mw Fort St. Vrain plant of Public Service of Colorado—was in pre-commercial "power ascension" phase as of mid-1976. Construction was substantially complete in early 1972 but significant technical problems—moisture in helium coolant, cracks in coolant system, and electrical cable redesign mandated by NRC after the Browns Ferry cable fire—have delayed commercial start-up by four years. This delay led to order cancellations by U.S. utilities during the financing squeeze of 1974-75.

General Atomic built a 40-Mw HTGR for Philadelphia Electric (Peach Bottom 1) in the late 1950's. It was decommissioned in 1974.

**Trend equations linking capacity factor to unit size in Sections 3.3 and 3.4 show about a third of the 9% difference between Westinghouse and GE units to be statistically attributable to greater average unit size of GE reactors. Also, Westinghouse reactors have had higher capacity factors than GE's for small and medium-size nuclear units (under 800 Mw capacity), but not at higher size levels.

***Duke's seven coal-fired units in the CEP coal data base have operated at an average capacity factor of 82.7% over 48 unit-years of operation, significantly higher than the utility average.

****In the first six months of 1976, three of five B&W reactors were shut down for refueling; a fourth underwent lengthy maintenance outage. A sixth reactor—the new Rancho Seco unit of Sacramento Municipal Utility District—has been shut down for most of its brief commercial history to repair steam leaks in the turbine. As a result, B&W reactors operated at 40.0% average capacity factor for the first half of 1976. Cumulative B&W reactor average was down to 56.8%.

TABLE 3.7: DISTRIBUTION OF PWR ONE-YEAR CAPACITY FACTORS

Capacity Factor	Number of Unit-Years	Number of Unit-Years, Discarding 1st Full Year
0- 9.9%	1	1
10-19.9%	0	0
20-29.9%	1	0
30-39.9%	6	2
40-49.9%	6	4
50-59.9%	11	5
60-69.9%	19	11
70-79.9%	15	12
80-86.4%	7	7
TOTAL	66	42

The low average capacity factor of Combustion Engineering reactors is due in large part to poor performance of the Palisades plant (23.2% capacity factor over four years of operation). Excluding Palisades, CE's two other reactors have a 54.3% average capacity factor. While this average reflects only four unit-years, CE reactors have generally operated well in the first half of 1976.*

3.2d: Capacity Factor and Unit Parameters

Nuclear capacity factors vary as a function of plant ownership, reactor type, and certain operating characteristics or "parameters." We have tested capacity factor against the following parameters: unit size, unit age, and year of initial commercial operation.**

Two or more parameters may vary together. For example, younger plants tend also to be larger than their elders. Required is a statistical tool that deduces for each parameter, or variable, its share of changes in capacity factor that result from changes in other or all variables. "Correlation analysis" is such a tool. Correlation analysis (referred to by economists as "regression analysis") relates variation in values of a dependent variable (capacity factor) to the independent effect of explanatory variables (unit size, age, and so forth) upon the dependent variable. Correlation analysis is not applicable where two or more explanatory variables are very closely linked (e.g., if all capacity factor observations for large units were based on ages one to two, while all observations for small units were based on ages five to eight). Fortunately that is not the case for the nuclear data base. Correlation equations were developed separately for PWR's and BWR's.

*CE reactors, including the new Calvert Cliffs 1 and Millstone 2 units, had a 61.7% average capacity factor during the first half of 1976. Cumulative capacity factor through June 1976 was 44.2%. Excluding Palisades, comparable figures are 72.9% (six months 1976) and 60.4% (cumulative through mid-1976) (see Appendix D).

**We have also tested effects of unit "duplication" (see Chapter 5).

Section 3.3: Pressurized Water Reactors—Statistical Findings

The 24 PWR's in the CEP data base have accumulated 66 full calendar-years of operation, with an average capacity factor of 61.6% and a standard deviation of 16.8%.

The range in PWR capacity factors is shown in Table 3.7 which indicates that the PWR distribution is centered in the 60%-70% range, absent eight capacity factors under 40%. Palisades registered four of the eight low capacity factors (including the two below 30%) due to steam generator problems. The other four are due to poor first full calendar years of operation by Prairie Island 1, San Onofre 1, Surry 1, and Zion 1.

The 24 PWR's were installed during 1968-1974. They range in capacity from 450 Mw to 1050 Mw, and in unit age from one to eight. Some differences in capacity factors are related to variation in the values of these variables. PWR capacity factor variation in the data base is best described by Equation 3.1.

*(Equation 3.1)**

$$\text{Capacity Factor} = 93.48 - \underset{(2.36)}{4.22} \times \frac{1}{\text{Log}(1 + \text{AGE})} - \underset{(1.14)}{3.368} \times (\text{Mw}/100)$$

Adjusted $R^2 = 0.21$
N = 66

Equation 3.1 shows PWR capacity factors have been related to *age* value of the reactor in a particular year, and to *unit size* (Mw capacity).

For example, expected capacity factor for an 800-Mw PWR in its fourth full year of commercial operation, based on PWR experience to date, could be estimated by substituting 800 for "Mw" and 4 for "AGE" in Equation 3.1.

$$\text{Or, Capacity Factor} = 93.48 - \frac{4.22}{\text{Log}(1 + 4)} - 3.368 \times \frac{800}{100},$$

$$\text{or, Capacity Factor} = 93.48 - \frac{4.22}{\text{Log } 5} - 3.368 \times 8,$$

$$\text{or, Capacity Factor} = 93.48 - 6.04 - 26.94,$$

$$\text{or, Capacity Factor} = 60.50\ (\%).$$

Equation 3.1 gives the best fit to actual PWR capacity factor data.

*The adjusted R-squared value measures the "goodness of fit" of equation to actual data. A value of zero indicates a random fit. A value of one denotes that each point is *on* the curve of the equation, (i.e., a perfect fit). Numbers in parentheses below coefficients of the variables are "standard error coefficients." A high ratio of coefficient to standard error coefficient indicates that the coefficient of that variable is probably a good approximation of the "actual" value of the variable. We have used a cut-off ratio of approximately 1.68 to eliminate variables with coefficients that have less than 95% statistical confidence. In Equation 3.1, the ratio of coefficient to standard error is 1.79 for the second term (4.22/2.36) and 2.95 for the third term (3.368/1.14).

N is the number of observations (unit-years) in the sample (see Appendix F).

TABLE 3.8: PWR CAPACITY FACTORS BY SIZE GROUPS

Size Group	Average Capacity Factor	# Units	# Unit-Years	Average Number of Years Per Unit
450-599 Mw	68.4%	9	34	3.8
600-799 Mw	62.4%	4	12	3.0
800-999 Mw	49.8%	9	17	1.9
1000-1050 Mw	47.9%	2	3	1.5
450-1050 Mw	61.6%	24	66	2.8

Year of commercial installation has not been statistically related to PWR capacity factor, as denoted by its absence as a variable from the right-hand side of the equation.* This absence of a "vintage effect"—a relationship between capacity factor and date of installation—indicates that design and construction of PWR's have not exhibited a capacity factor "learning curve" during the commercial service entry period of 1968-74. This is no surprise, however, in that units brought into service by 1974 were generally well into design or initial fabrication stages in 1968. Thus, there was limited opportunity to feed back into newer units the results of operating experience.** Increased safety-related regulatory requirements and rapid scaling up of unit sizes by reactor vendors have also precluded design duplication and stability which foster learning.***

Ramifications of Equation 3.1

Unit Size: The term, "$-3.368 \times (\text{Mw}/100)$," indicates that PWR capacity factors to date have declined by an average of 3.368 percentage points per 100 Mw increase in capacity, within the size range contained in the data base (450 to 1050 Mw), independent of the effect of age. This indicates that superior capacity factors of smaller over larger PWR's are largely intrinsic to unit size, not just a result of the greater "maturity" of smaller PWR's.

PWR capacity factors fall steadily for each higher size group as indicated by Table 3.8. Decline appears to average just over 4 percentage points per 100 Mw increase in capacity. Comparison with the coefficient (negative 3.368/100 Mw) in Equation 3.1 implies that less than a fifth of the drop per 100 Mw is due to shorter average age of larger PWR's. The remainder is an intrinsic function of size.

Unit Age: Unit age and PWR capacity factor are related in Equation 3.1 by the term, "$-4.22 \times [1/\text{Log}(1 + \text{AGE})]$". The 1/Log term decreases as AGE increases, so the negative sign of the coefficient (-4.22) indicates PWR

*That is, the standard error term of the coefficient of year of commercial installation in a correlation equation involving age, size, and year of installation, was too *great* to impart confidence at the 95% level to the coefficient. The correlation equation, along with other "rejected" equations, is shown in Appendix E.

**However, higher capacity factors of "duplicate" PWR's suggest improvements need not require a long learning period (see Chapter 5).

***We are indebted for these observations to Evan C. Kovacic, Director, Power-plant Productivity Division, Office of Fuels and Facilities, FEA (private communication, August 13, 1976).

TABLE 3.9: PWR UNIT-YEAR CAPACITY FACTORS BY AGE, FOR DIFFERENT SIZE GROUPS

Size Group (in Mw)	AGES 1-2 Average Capacity Factor	AGES 1-2 Number of Unit-Years	AGES 3+ Average Capacity Factor	AGES 3+ Number of Unit-Years	Greatest Age in Sample
450-599	62.0%	15	73.4%	19	8
600-799	59.0%	8	69.3%	4	4
800-1050	53.0%	17	29.7%	3	4
800-1050 w.o. Palisades	56.2%	15	54.3%	1	3
All PWR's	57.6%	40	67.7%	26	8

capacity factor to be *positively* correlated with unit age (*i.e.*, capacity factor increases as age increases).

A typical PWR would show about a 5% capacity factor increase in year two, a 2% increase in year three, a 1% increase in year four, levelling to minor increases (0.2%) by year ten.*

The impact of the age term upon capacity factor for PWR's is illustrated in Figure 3.1, for three different sizes, 600 Mw, 800 Mw, and 1000 Mw. As seen from the curves, a 10 percentage point improvement in capacity factor between age one and age ten for any unit size is represented by the age term in Equation 3.1, with half of the improvement occurring as the unit advances from age one to age two.**

Examination of actual PWR experience indicates those under 800 Mw capacity exhibit greater maturation than is indicated by the age term. Conversely, large PWR's operate at *reduced* capacity factors after two full commercial years. However, the decline in large PWR capacity factors is attributable to Palisades. Excluding Palisades, there is only one unit-year of experience for 800+ Mw PWR's after age two.

*For the reader's convenience, we present a table giving the value of 1/Log (1 + AGE) and the full AGE term in Equation 3.1 [4.22 × 1/Log (1 + AGE)] for different values of AGE:

AGE	Log Function	Term in Equation 3.1
1	3.32	14.02
2	2.10	8.84
3	1.66	7.01
4	1.43	6.04
5	1.29	5.42
6	1.18	4.99
7	1.11	4.67
8	1.05	4.42
9	1.00	4.22
10	0.96	4.05

**A precise relationship between capacity factor and unit age for PWR's is *not* necessarily provided by the log function discussed here. Rather, the log function provides the best "fit" for describing the *average* relationship between capacity factor and unit age. Some PWR's show greater maturity than indicated by Equation 3.1, while others show decreased capacity factors with age. Moreover, the two oldest PWR's in the CEP data base each have only eight years of operation. Projections beyond age eight (using Equation 3.1) are based strictly on extrapolation. See Chapter 5 for evidence of possibly *declining* capacity factors after age ten, rather than constant.

FIGURE 3.1

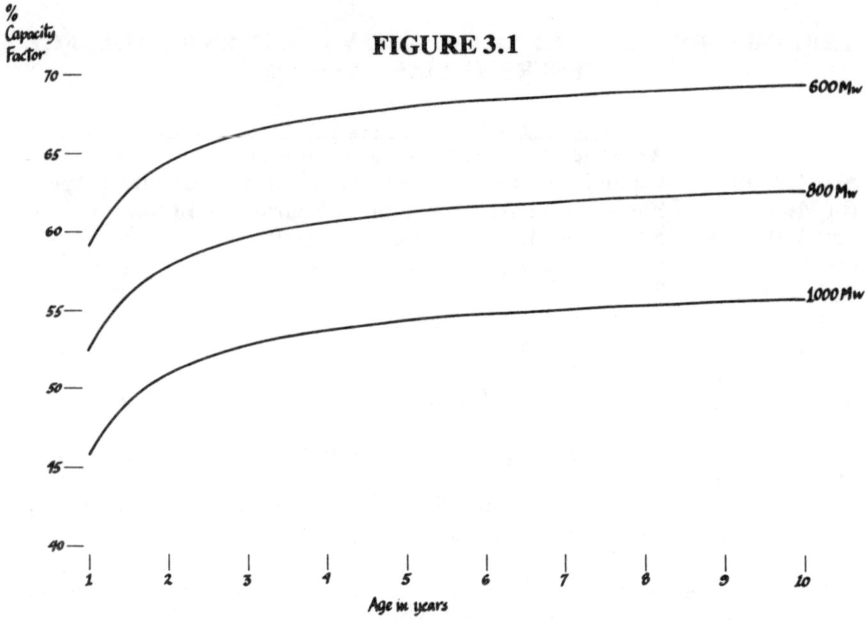

Table 3.9 indicates an inverse relation of capacity factor to unit size for different unit ages. This further validates the index of capacity factor penalty with increased size indicated by Equation 3.1. Other PWR regression equations yield no statistically significant relation of capacity factor to unit size for AGE = 1. But they do produce capacity factor "penalties" of 2.989 percentage points per 100 Mw increment for AGE = 2, and 7.920 points per 100 Mw for AGE = 3+ (see Appendix E).

Using Equation 3.1: Equation 3.1 may be used to project PWR capacity factors for various unit sizes and ages, using only figures to date as a projection base. Since economic analysis of generating units is influenced more by the projected *average* capacity factor of a unit over time than that in a given year, it is useful to convert Equation 3.1 accordingly. We have chosen the first 10 years of commercial operation. Nuclear plants are nominally projected to have 30-year lives, but PWR experience in the CEP data base is only through age eight. Absent Connecticut Yankee and San Onofre 1, it is only through age five. Choice of a 10-year period is a compromise between a long planning horizon and a short operating history. Equation 3.2, computed from Equation 3.1, provides the average capacity factor for ages one to ten of a PWR as a function of unit size.

*(Equation 3.2)**

$$\text{Capacity Factor} = 86.38 - 3.368 \times (\text{Mw}/100)$$

*This is a "levelized average," computed using a 10% discount rate. A levelized average gives greater weight to earlier capacity factors than to later (reflecting greater value of constant amounts of dollars, kwh, etc. received in early years), and so is slightly less than an arithmetic average for PWR capacity factors. Based on an arithmetic average of capacity factors in years one to ten Equation 3.1 would become Capacity Factor = 87.11 − 3.368 (Mw/100) (see Appendix F).

FIGURE 3.2

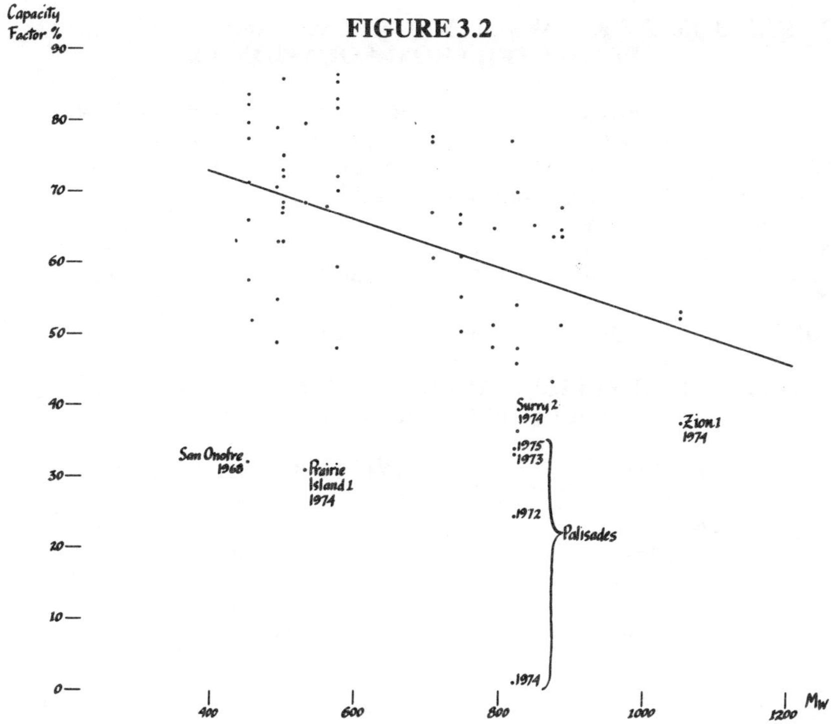

Figure 3.2 graphs the linear relationship between unit size and capacity factor given by Equation 3.2 and adds a "scatter plot" of unit size and capacity factor of the 66 PWR unit-years in the CEP data base. One-year capacity factors under 40% are identified in Figure 3.2.

Table 3.10, based on Equation 3.2, projects PWR average capacity factors for ages one to ten and compares them to actual PWR experience to date. Thus it is clear that actual and projected (levelized) capacity factors for small and medium-size PWR's are about equal. But projected levelized capacity factor for large PWR's is greater than the experience to date. In effect, Equations 3.1 and 3.2 attribute to large PWR's maturation that has occurred to date in small and medium PWR's.*

It is speculative to assume that large PWR's will improve in capacity factor at a rate commensurate with aggregate PWR experience to date. Large PWR's may mature little or not at all. Conversely, after their early years, they may exhibit greater improvements in capacity factor than smaller units, especially so if, as many in the nuclear industry believe, a longer "learning period" is required for large units, or at least the initial large units.

Despite projective limitations, to the extent Equations 3.1 and 3.2 summarize PWR capacity factor data to date (with size and age trends therein), they are a

*The equations attribute average maturation that has occurred in *all* PWR's to date, which was actually contributed by PWR's 450-799 Mw and was not observed in the limited experience available from PWR's 800 Mw or larger.

TABLE 3.10: PWR CAPACITY FACTORS, AGE 1-10 AVERAGE, PROJECTED FROM EQUATION 3.2

Size Group (in Mw)	Average Capacity in CEP Data Base	Average Age in CEP Data Base	Average Capacity Factor in CEP Data Base	Levelized Capacity Factor Projected for Capacity in Column 2
450-599	507 Mw	3.4	68.4%	69.3%
600-799	744 Mw	2.1	62.4%	61.3%
800-1050	875 Mw	1.7	49.5%	56.9%
All PWR's	662 Mw	2.6	61.6%	64.1%

TABLE 3.11: LEVELIZED AVERAGE PWR CAPACITY FACTORS, AGES 1-10, FROM EQUATION 3.2

Unit Size	Projected Capacity Factor
500 Mw	69.5%
600 Mw	66.2%
700 Mw	62.8%
800 Mw	59.4%
900 Mw	56.1%
1000 Mw	52.7%
1150 Mw	47.6%
1300 Mw	42.6%

TABLE 3.12: DISTRIBUTION OF BWR ONE-YEAR CAPACITY FACTORS

Capacity Factor	Number of Unit-Years	Number of Unit-Years Discarding 1st Full Year
10-19.9%	1	0
20-29.9%	0	0
30-39.9%	5	4
40-49.9%	5	4
50-59.9%	12	8
60-69.9%	16	12
70-79.1%	5	2

TABLE 3.13: BWR CAPACITY FACTORS BY SIZE GROUPS

Size Group (in Mw)	Average Capacity Factor	# Units	# Unit-Years	Average Number of Years Per Unit
514-599	62.7%	2	7	3.5
600-799	56.6%	5	21	4.2
800-1098	52.1%	7	16	2.3
514-1098	55.9%	14	44	3.1

starting point.* Table 3.11 projects average (levelized) capacity factors for years one to ten, for various PWR unit sizes, based on Equation 3.2.

Section 3.4: Boiling Water Reactors—Statistical Findings

The 14 BWR's in the CEP data base have accumulated 44 full calendar-years of operation, with an average capacity factor of 55.9% and a standard deviation of 14.1%. The range in BWR capacity factors is shown in Table 3.12. Absent the six capacity factors under 40%, the median BWR distribution is around 60%. In comparison, the PWR median is in the mid-to-high 60%-70% range, also absent sub-40% capacity factors. Whereas the eight very low PWR capacity factor were attributable to the Palisades plant or first full year operation of other units, the sub-40% BWR capacity factors follow no discernible pattern. Two occur at age one (Browns Ferry 1, Nine Mile Point 1); one at age two (Pilgrim); two at age three (Millstone 1, Quad Cities 2); and one at age four (Dresden 3). Moreover, comparison of unit-year columns in Table 3.12 indicates that first calendar years of operation account for many BWR high-capacity factor years and few low-capacity.

Table 3.12 also indicates that no BWR has operated at an 80% capacity factor for any of the 44 unit-years accumulated to date (whereas PWR's have seven of 66 unit-years above 80%), despite AEC and industry predictions to the contrary. The highest BWR capacity factor in any single calendar year was 79.1% (1975) by Vermont Yankee, which is the smallest capacity BWR (514 Mw).

The 14 BWR's were installed during the years 1969-74, and range in capacity from 514 Mw to 1098 Mw, and in unit age from one to six.

Of the three variables (unit size, unit age, and year of commercial installation), only unit size is related to BWR capacity factor in a statistically significant way. This relationship is given in Equation 3.3

(Equation 3.3)

$$\text{Capacity Factor} = 79.30 - \underset{(1.46)}{3.287} \times (\text{Mw}/100)$$

$$\text{Adjusted } R^2 = 0.11$$
$$N = 44$$

According to Equation 3.3 an average capacity factor decline of 3.287 percentage points is associated with each BWR capacity increment of 100 Mw. This is almost identical to the coefficient for PWR's of minus 3.368 points for each 100 Mw increment (Equations 3.1 and 3.2). However, unit age and year of

*Equations 3.1 and 3.2 are based on PWR operation through the end of 1975. We have also developed correlation equations for PWR capacity factors, based on operation through mid-1976, by adding seasonally adjusted data from the first half of 1976 to the data base. These equations project lower capacity factors for PWR's above 676 Mw capacity, as compared to Equations 3.1 and 3.2 (see Appendix D).

TABLE 3.14: BWR UNIT-YEAR CAPACITY FACTORS BY AGE, FOR DIFFERENT SIZE GROUPS

	AGES 1-2		AGES 3+		
Size Group (in Mw)	Average Capacity Factor	Number of Unit-Years	Average Capacity Factor	Number of Unit-Years	Greatest Age in Sample
514-599	59.6%	4	66.9%	3	4
600-799	54.0%	9	58.6%	12	6
800-1098	56.5%	11	42.6%	5	4
All BWR's	56.1%	24	55.8%	20	6

installation were not statistically related to capacity factor.*

As with PWR's, the term, "$-3.287 \times$ (Mw/100)," (and absence of a term for unit age) suggests that current superior performance of smaller over larger BWR's results from intrinsic factors related to unit size, rather than greater maturity of smaller BWR's. This is apparent in Table 3.13. BWR capacity factors fall steadily for each higher size group, notwithstanding slightly greater average experience with medium-size (600-799 Mw) BWR's.

Examination of BWR experience disaggregated by unit size and age indicates those under 800 Mw have improved modestly with increased age, but large BWR's have suffered reduced capacity factor after age two. This is similar to PWR's. Unlike PWR's, however, age improvement in capacity factor of sub-800 Mw BWR's did not offset age decline for large BWR's enough to produce an aggregate maturation trend.

The bottom line of Table 3.14 demonstrates that average BWR capacity factor after age two has equalled that up through age two. The table also shows that large BWR's performed on the same level as smaller BWR's during ages one to two, but at a significantly lower level after age two. The same conclusion is reached by regression analysis of different BWR age groups. No statistically significant relationship was found between capacity factor and unit size for "AGE = 1" or "AGE = 2," but a capacity factor "penalty" of 9.614 percentage points per 100 Mw increment was obtained for "AGE = 3+" (see Appendix E). The coefficient ($-3.287/100$ Mw) in Equation 3.3 is in effect a result of the significant difference between capacity factors for large and smaller BWR's after age two.

Figure 3.3 graphs the linear relationship between unit size and capacity factor given by Equation 3.3, and adds a "scatter plot" of the unit size and capacity factor of each of the 44 BWR unit-years in the CEP data base.

Table 3.15 indicates a close agreement between actual and projected capacity factors for BWR's. Future BWR performance will not necessarily mirror performance to date, and only a small percentage of variation in BWR capacity factor between years and units is explained by Equation 3.3. But the equation is

*Because of the heavy representation of Commonwealth Edison in the BWR sample—13 out of the 44 BWR unit-years are accounted for by Commonwealth Edison's Dresden 2 and 3 and Quad Cities 1 and 2—we also tested (in regression equations) to determine whether Commonwealth Edison ownership was correlated with capacity factor. It was not. The average capacity factor of 54.5% of Commonwealth Edison's BWR's, all of 809 Mw capacity, is close to the capacity factor which would be predicted from Equation 3.3 for BWR's of that capacity (see Appendix E).

FIGURE 3.3

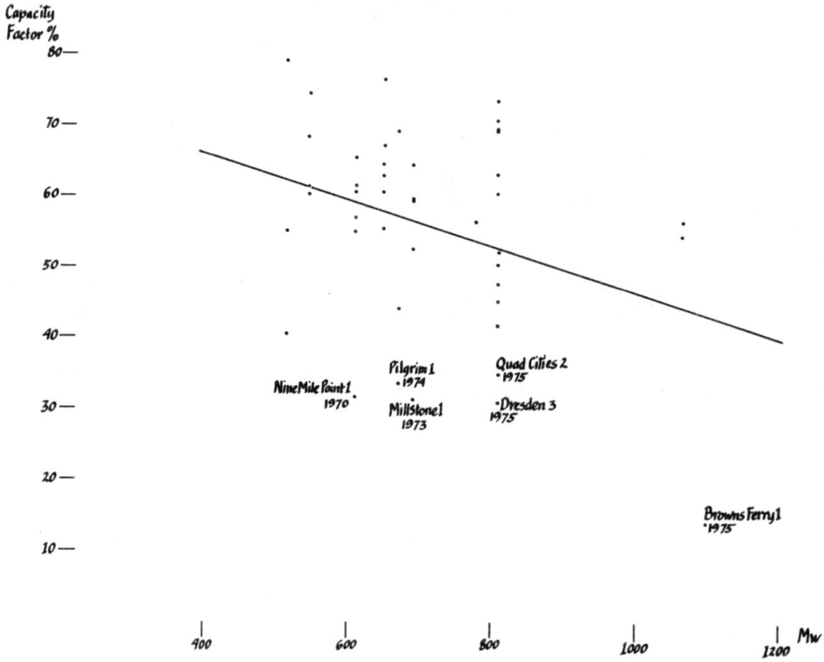

a starting point for projecting future BWR capacity factors. Table 3.16 projects BWR capacity factors for different unit sizes.*

BWR-PWR Comparison: BWR and PWR capacity factors to date, and those projected from Equations 3.2 and 3.3 are summarized in Table 3.17.

Below 800 Mw, actual and projected PWR capacity factors are both roughly six percent higher than BWR historical and projected capacity factors. Above 800 Mw, PWR capacity factors averaged 2½ percentage points below BWR capacity factors. But PWR capacity factors are projected to be roughly six points higher, based on Equations 3.2 and 3.3. The projected superiority of large PWR's over large BWR's is based on the assumption that large PWR's will display the same maturation (improvement in capacity factor with increasing age) shown by all PWR's.

*Equation 3.3 is based on BWR operation through the end of 1975. We have also developed a correlation equation for BWR capacity factors based on operation through mid-1976, by adding seasonally adjusted data from the first half of 1976 to the data base. This equation projects significantly lower capacity factors for BWR's above 697 Mw capacity, as compared to Equation 3.3. The new equation also has a significantly better goodness of fit than Equation 3.3 (see Appendix D).

TABLE 3.15: BWR CAPACITY FACTORS PROJECTED FROM EQUATION 3.3 †

Size Group (in Mw)	Average Capacity in CEP Data Base	Average Capacity Factor in CEP Data Base	Capacity Factor Projected for Capacity in Column 2
514-599	532 Mw	62.7%	61.8%
600-799	657 Mw	56.6%	57.7%
800-1098	859 Mw	52.1%	51.1%
1000+ only	1076 Mw	41.7%	43.9%
All BWR's	711 Mw	55.9%	55.9%

†We have dispensed with the formulation of an average capacity factor for ages 1-10 (used for PWR's), because the absence of an age term in Equation 3.3 implies that BWR capacity factors will be constant over time.

TABLE 3.16: BWR CAPACITY FACTORS, FROM EQUATION 3.3

Unit Size	Projected Capacity Factor
500 Mw	62.9%
600 Mw	59.6%
700 Mw	56.3%
800 Mw	53.0%
900 Mw	49.7%
1000 Mw	46.4%
1150 Mw	41.5%
1300 Mw	36.6%

TABLE 3.17: BWR & PWR CAPACITY FACTORS—HISTORICAL & PROJECTED

	Historical				10-Year Projected		
	BWR		PWR				
Size Group (in Mw)	Average Capacity (in Mw)	Average Capacity Factor	Average Capacity (in Mw)	Average Capacity Factor	Capacity (in Mw)	Capacity Factor BWR/PWR	
450-599	532	62.7%	507	68.4%	500	62.9%	69.5%
600-799	657	56.6%	744	62.4%	700	56.3%	62.8%
800-1098	859	52.1%	875	49.5%	900	49.7%	56.1%
1000-1098	1076	41.7%	1050	47.9%	1100	43.1%	49.3%

4: Factors Affecting Nuclear Plant Capacity Factors

Section 4.1: Key Findings

1. The average 40.7% "lost" capacity factor at a typical nuclear plant (based on an average 59.3% capacity factor) is accounted for as follows:

a. *Scheduled outages,* which comprise refueling every 12-18 months and maintenance have been responsible for 18% in lost capacity factor at an average nuclear plant. Maintenance refers to periodic shutdowns required to maintain, repair, and test equipment. This is two to three times the anticipated capacity factor loss in optimistic industry projections of refueling and maintenance requirements.

b. *Equipment malfunctions,* which are failures in equipment causing power reduction or complete shutdown, have accounted for 17½% capacity factor loss. Of this total, 10½% has been caused by plant shutdowns for equipment failure (forced outages), while 7% has been associated with partial power reductions attributable to malfunctioning equipment and faulty nuclear fuel (partial forced outages).

c. *Regulatory restrictions,* such as NRC-mandated inspections or power de-ratings to avoid violating license specifications, have caused 4% in lost capacity factor.

d. *Load-following and time to come to full power,* are deliberate power reductions due to low system electric demand, and mechanical and regulatory limits on the rate of power increase following shutdown. They have together accounted for 1% in lost capacity factor.

2. PWR's have been troubled most frequently by equipment failures (such as leaks) in turbines and steam generators. BWR's have been plagued by failures of nuclear fuel (such as leaks in fuel rod cladding) and of reactor components (such as coolant water circulating systems) which have lengthened outages for refueling and maintenance.

3. No single cause dominates nuclear capacity problems. For instance, in equipment malfunction loss the largest single category is fuel failures, yet they

account for only 3.5% of lost capacity factor. Another 5.2% are so varied as to defy classification into a single category. Similarly, no one cause accounts for the significant capacity factor loss attributable to refueling and maintenance.*

Section 4.2: Refueling and Scheduled Outages

Nuclear plants need to be "refueled" at intervals so that "spent" fuel elements (containing uranium oxide fuel plus neutron-absorbing elements produced by nuclear fission) in the reactor core may be removed and replaced by "fresh" fuel. Reactor cores consist of several "fuel regions" characterized by different residence periods in the reactor. A fuel region is usually removed after about four years in the core—when its component fuel elements have undergone sufficient fission reactions so that their residual fuel is insufficiently contributing to the total nuclear fissions in the reactor core.

Reactors are designed to be refueled at approximate 12-month intervals, based upon 80% capacity factors. The first refueling, however, is generally scheduled 18 months after start-up, since all of the initial fuel is fresh and can be "irradiated" for longer periods without losing "reactivity." In practice, refueling has occurred less frequently at many U.S. nuclear plants because of lower capacity factors (fuel irradiation is proportional to capacity factor, so that lower capacity factors imply reduced irradiation levels). The reduction in refueling frequency, however, may not be directly proportional to the decrease in capacity factor because a plant near the end of a year's "fuel cycle" would commonly be refueled ahead of schedule when it suffers an extensive outage to repair or maintain other equipment. This is because the value of unused fuel is typically less than the cost of a separate downtime.

In other cases, refueling has occurred early to replace "defective" fuel which could otherwise cause undesirable levels of radioactivity in the reactor, and to ensure that plants are not operated with insufficient control rod "worths." The ability of boron control rods to absorb fission-inducing neu-

*Estimates of nuclear plant capacity reductions are based on a recent report by the Electric Power Research Institute (EPRI).[1] In it is a comprehensive review of causes of nuclear plant outages, covering the period January 1973 through June 1975. Close to 60% of the nuclear unit-years in the CEP data base were logged in the period studied by EPRI.

The report estimates refueling outages have averaged 1500 hours in length at nuclear plants; that refuelings have occurred at average intervals of one year; and that all other outages have accounted for an average of 1486 hours of down-time between refueling. Nuclear plants have been "available" to operate 71% of the time [100% − (1500 + 1486)/(8760 + 1500)]. They have operated at an average capacity factor of 85% during available hours for an average annual capacity factor of 60.4% (71% of 85%).

From 1973-75, nuclear units in the CEP data base operated at an average capacity factor of 57.8%—less than EPRI's 60.4%, but close enough to be consistent with EPRI's figures.

Percentage estimates of capacity factor lost to each cause category are computed as the sum of:

1) Full outage time, which is the product of the percentage of total outage time attributed to each category by EPRI times 0.407 (one minus the average 1968-75 CEP nuclear capacity factor of 59.3%).

2) Equivalent capacity factor lost to partial power reductions, which is the product of the reduction in "output factor" for each category multiplied by EPRI's "availability factor" average of 0.71.

Unless otherwise noted, estimates of capacity factor reductions are computed by this method from EPRI figures.

trons decreases as neutrons are absorbed by the rods over time. Refueling may also be advanced or postponed to schedule the outage during a "load window"—a season of low electricity demand for the utility system.

"Refueling outage" has actually come to be used in the nuclear industry as an umbrella term under which many annual maintenance and inspection activities are performed. This was noted by General Electric engineers in a 1974 paper, "there is a considerable amount of maintenance that can be scheduled only concurrently with a major plant shutdown."[2] This includes major maintenance of the turbine-generator and any work on equipment located within the primary containment. Such work can include repair of primary system relief valves, safety valves, isolation valves, control rod drives, recirculation pumps, in-core instrumentation, and much of the in-service inspection work required under Section 11 of the boiler and pressure vessel code of the American Society of Mechanical Engineers.*

The available data on refueling indicates that the average *duration* of refueling outages has significantly exceeded the initial expectations of nuclear plant designers and operators. An EPRI survey of 30 U.S. nuclear units through March 1, 1975, indicates average refueling durations for plants over 100 Mw to be 12.1 weeks for BWR's and 10.0 weeks for PWR's.[4]

Assuming a 15-18 month interval between refuelings, the EPRI figures correspond to an average annual capacity factor reduction of 11%-13% for PWR's and 13%-16% for BWR's. Optimistic industry estimates set refueling outage time at three to four weeks, equivalent to a 5%-7% capacity factor loss, assuming annual refueling. Industry estimates are based on a "critical path" model which assumes refueling to be the "pacing" activity in a refueling outage with no problems discovered to extend the outage time.

Utilities have begun to acknowledge, however, that refueling experience has generally not conformed to this model. For example, a study for the Federal Energy Administration reported that Duke Power and Jersey Central Power & Light have recently increased their estimates of future refueling outage duration at their Oconee (PWR) and Oyster Creek (BWR) plants, respectively, from four weeks to six weeks. This was based on a reappraisal of scheduled maintenance requirements other than refueling the core.[5]

Recent refueling at these plants has exceeded even this revised projection. Oyster Creek's most recent refueling outage was projected at eight weeks and was to include turbine inspection, retube of the main condenser, and modification of the emergency core cooling system. It actually required 10½ weeks. Similarly, Oconee 1 was down for 8½ weeks for its recent refueling.**

*For example, Commonwealth Edison reported that the Spring 1976 refueling at Dresden 2 would entail (besides fuel unloading, reloading, and shuffling) the following: feedwater sparger and jet pump inspections, control rod drive overhaul, in-service inspections, local leak rate tests, local power range monitor replacement, and integrated drywell leak rate test, plus turbine work. The refueling outage commenced March 15, 1976, and was concluded May 27.[3] The 74 day outage time is less than average duration of BWR refueling outages to date.

**Oconee 2 was shut down on April 7, 1976, for reactor core inspection and repair work and began refueling on May 15. As this study went to press, the scheduled return date was July 11. At best, Oconee 2's refueling will have required eight weeks, with 13½ weeks for the complete maintenance and refueling outage. Oconee 3 has not undergone refueling since the FEA report was prepared.[6]

Southern California Edison, whose San Onofre unit has operated at a relatively high 68.8% capacity factor over its eight year life, reports that reactor plant refueling and maintenance activities have usually required only 30 days. But steam generator inspections and repairs, turbine overhauls and maintenance have usually required total outages to be extended to 50-55 days.[7] San Onofre's Spring 1976 refueling required 40 days.

General Electric argues that refueling plus "normally anticipated annual maintenance" could be done in three weeks, based on a critical path model. But it concedes that "unusual maintenance in other parts of the plant, such as a complete turbine inspection, could extend the refueling and maintenance outage beyond that length of time."[8]

United States' BWR's, manufactured by GE, have had particularly long refueling outages during 1975 and early 1976. Principal maintenance was to replace failed feedwater spargers (circular pipe distributing devices which "feed" a continuous flow of water to the reactor core), to inspect and in some cases replace recirculating systems (which circulate water in the core), to inspect for possible cracks in by-pass lines (and repair welds), to modify local power range monitors (devices which meausre neutron flux), and because of fuel damage which has required lengthy inspection of fuel rods and has complicated in-core work by increasing radiation levels. These problems have required extensive maintenance during refueling at Commonwealth Edison's four BWR's (see Table 4.1).

The recent refueling and maintenance outage at Boston Edison's BWR Pilgrim 1 unit illustrates difficulties in accurate projection of refueling outage time and in ways to minimize it. Pilgrim 1 was shut down January 27, 1976, for its first major refueling.* It had a projected start-up date of April 1. The plant actually returned to service on June 4, over four months after shutdown and two months behind schedule. At least half of the delay was caused by three major equipment problems unanticipated prior to shutdown: (1) the feedwater sparger had to be removed and replaced and superficial cracks in the sparger nozzles had to be ground out (two weeks); (2) bolts that secure the riser on the jet pump, which helps circulate primary coolant water in the reactor, had to be replaced (one week); (3) and refueling itself was extended (two weeks) by mechanical and electrical problems with the refueling bridge—a mechanized structure above the reactor core used to load and unload fuel. Minor problems with water chemistry, hydrotesting, and so on, accounted for the remainder of the extension.[9]

Conversely, Point Beach 2 completed its second refueling outage, early this year, in only 29 days. It is a PWR which has averaged an exceptionally high 75.9% capacity factor during three years of commercial operation. Lack of added major maintenance work is credited for the short duration. Similarly, Monticello was able to refuel in only 28 days in 1975. Its 66.2% capacity factor over four years is the highest lifetime BWR capacity factor to date.

A complete plant-by-plant record of refueling outage duration and frequency is not readily available. Table 4.1 includes all refueling and maintenance

*A small number of defective fuel bundles had been replaced in a minor fuel change in early 1975.

TABLE 4.1: NUCLEAR PLANT REFUELING AND MAINTENANCE OUTAGES JANUARY 1975-JUNE 1976

Unit	Date Shut Down	Date Returned	Duration Days	# of Times Refueled
Pressurized Water Reactors				
Connecticut Yankee	5-17-75	7-1-75	45	5
San Onofre 1	3-14-75	4-23-75	40	4
Ginna	1-29-76	4-10-76	72	5
	3-10-75	5-19-75	70	4
Point Beach 1	11-16-75	1-8-76	53	3
Palisades	12-20-75	5-9-76	141	2
Robinson 2	10-31-75	12-11-75	42	2
Maine Yankee	5-2-75	6-27-75	56	2
Point Beach 2	2-26-76	3-26-76	29	2
Turkey Point 3	10-25-75	12-24-75	60	2
Surry 1	10-1-75	12-8-75	68	2
	10-24-74	2-3-75	102	1
Oconee 1	2-7-76	4-6-76	59	2
	1-19-75	3-13-75	53	1
Surry 2	4-22-76	6-9-76	48	2
	4-25-75	6-17-75	53	1
Turkey Point 4	4-18-76	6-10-76	53	2
	3-29-75	6-21-75	84	1
Prairie Island 1	3-4-76	5-4-76	61	1
Fort Calhoun	2-7-75	5-9-75	91	1
Zion 1	3-5-76	6-19-76	106	1
Three Mile Island	2-20-76	5-27-76	97	1
Indian Point 2	3-30-76	—		1
Oconee 2	4-7-76	—		1
Oconee 3		(to be refueled 9-1-76)		
Arkansas 1		(to be refueled 10-1-76)		
Prairie Island 2				
Zion 2				
Kewaunee				
Boiling Water Reactors				
Oyster Creek	12-27-75	3-10-76	74	5
	3-29-75	5-24-75	56	4
Nine Mile Point 1	9-13-75	12-4-75	82	4
Millstone 1	9-13-75	10-20-75	37	4
Dresden 2	3-14-76	5-27-76	74	4
	11-2-74	5-18-75	197	3
Monticello	9-11-75	11-19-75	69	3
	1-10-75	2-7-75	28	2
Dresden 3	4-16-75	9-10-75	147	2
Quad Cities 1	1-3-76	3-14-76	71	2
Quad Cities 2	12-23-74	4-26-75	124	2
Pilgrim 1	1-27-76	6-4-76	129	1
Vermont Yankee	(refueled late 1974; to be refueled 6-18-76)			
Peach Bottom 2	3-27-76	6-24-76	89	1
Peach Bottom 3		(to be refueled 11-25-76)		
Browns Ferry 1	(shut down since March 1975 to repair fire damage)			
Cooper		(to be refueled 9-1-76)		

outages for the 18-month period January 1, 1975, through June 30, 1976, from the NRC Gray Books, arranged in approximate decreasing unit age.

Table 4.1 includes 22 refueling outages at 17 of 24 PWR's in the data base. Two PWR's had not completed their refueling outages during the survey period and five others awaited first refueling. It also includes 13 refueling outages at 10 of 14 BWR's. Three units awaited first refueling and one was refueled just before and just after the survey period.

Average PWR outage duration was 9.6 weeks and BWR's was 12.9 weeks. This is slightly less than average durations (through March 1, 1975) reported by EPRI for PWR's, slightly more for BWR's. Though conclusions are limited by an 18-month test period, it is notable that the 23-day difference between average BWR and PWR refueling outage duration in Table 4.1 is equivalent to a 5% capacity factor difference, assuming 15 months for a complete operating and refueling cycle. This almost equals the net 5.5% difference between average BWR and PWR capacity factors in the CEP data base.*

Due to increased plant operator proficiency and a decrease in inspection and maintenance requirements as plants are "broken in," nuclear engineers generally anticipate duration of refueling outages to diminish as plants mature. This expectation is reflected in GE's esimate that BWR refuelings will drop from eight to ten weeks for the first two refuelings to five to seven weeks for subsequent refuelings.**[11] Westinghouse has also asserted that a refueling "learning curve" (reduced outage durations) has been shown by individual plants. The data presented, however, is scarce beyond the second refueling.[13]

Some support for this view is provided by Table 4.1, which indicates that second and subsequent refueling outages have required less time than first outages.*** However, a more comprehensive survey is needed for substantiation. Moreover, especially lengthy recent outages for several units suggest that scheduled maintenance is considerably extending refueling outages, thereby obscuring the data.

Data is limited as to reasons for and durations of scheduled outages other than refueling. The EPRI cause classification of nuclear plant capacity reductions indicates that of 18% capacity factor lost at the average nuclear unit for scheduled shutdowns, 15% is attributable to refueling and other activities conducted under the umbrella of "refueling," and 3% to separate maintenance outages.****[14]

*Not included are the Oconee 2 and Indian Point 2 (PWR's) refueling and maintenance outages, which lasted at least three and five months, respectively, and were not complete by the end of the survey period. Moreover, there is no adjustment for *frequency* of refueling.

**These figures are less than EPRI's average duration of 12.1 weeks for 100+ Mw BWR's, and the 13.0 week average from Table 4.1, and may be explained in part by the possible inclusion of several older sub-100 Mw BWR's, which have had better refueling records.[12]

***From Table 4.1, average PWR refueling outage duration was 56.0 days for third and subsequent refuelings (five units), 61.8 days for second refueling (nine units), and 80.9 days for first refueling (eight units). For BWR's, average durations were 84.1 days for third + refuelings (six units), 92.5 days for second refueling (four units), and 109 days for the two first refuelings during the survey period.

****EPRI notes that, "Evaluations of scheduled outages are inhibited by both terminology and by limited accessibility to detailed evaluations of the nature of operations which occur in the period nominally referred to as refueling . . . In general, currently available reporting of refueling operations does not permit useful isolation of discrete maintenance and repair activities."[15]

TABLE 4.2: ESTIMATED AVERAGE NUCLEAR CAPACITY FACTOR REDUCTIONS DUE TO EQUIPMENT MALFUNCTIONS, JANUARY 1973-JUNE 1975

Equipment Responsible	Capacity Factor Lost Full Forced Outages†	Partial Forced Outages
Turbine-Generators	1.5%	
Steam Generators	1.9%	
Condensers	1.3%	1.4%
Valves	1.4%	
Pumps	0.6%	
Nuclear Steam Supply System (includes fuel failure)	0.7%	3.5%
Other	3.1%	2.1%
TOTAL	10.5%	7.0%

†Capacity factor lost due to full forced outages was computed by multiplying (1) percentage of total nuclear outage time accounted for by each item, Lapides & Zebroski, *op. cit.*, p. 16, multipled by (2) average nuclear unit ''unavailability'' of 29%, *ibid.*, times (3) ratio of EPRI approximate capacity factor of 60.4% to CEP average capacity factor of 59.3%, to ''normalize'' EPRI's outage estimates to the CEP data base. Example: Steam generators account for 6.3% of nuclear downtime, per EPRI, and are therefore responsible for a 1.9% average capacity factor loss (.063 × 0.29 × 0.604/0.593).

Capacity factor lost because of partial forced outages was computed from EPRI's estimate that nuclear units have averaged 71% ''availability'' (are operating at some power output 71% of the time), and EPRI's estimate of de-ratings while units were in operation *(ibid.)*.

Section 4.3: Equipment Malfunctions: Forced Outages and Partial Forced Outages

The EPRI survey suggests an average capacity factor loss of 17½% by U.S. nuclear units due to equipment malfunctions and associated downtime. About three-fifths of this reduction (10½%) is due to plant shutdowns for equipment failure (forced outages). The remaining two-fifths (7%) is associated with partial power reductions attributable to fuel failures and other equipment malfunction (partial forced outages). EPRI acknowledges that, ''Operating experience data covering unit partial power reductions are not readily available.''[16] However, EPRI's data does offer a rough cause classification for partial power reduction.

Close to half of the average 10.5% capacity factor loss associated with forced outages is accounted for by about 70 ''major incidents'' (outages of greater than 500 hour duration) in the January 1973-June 1975 period studied by EPRI. Remaining forced outage time is attributed to several thousand incidents in that period. Clearly, a significant reduction in major outages would improve nuclear capacity factors—by up to 5% based on the experience in the EPRI survey.

Table 4.2 uses EPRI data to estimate nuclear capacity factor lost under full forced outages and partial forced outages.

1. *Turbine-generators* (1.5% capacity factor loss): Virtually all turbine-generator sets at U.S. nuclear plants are supplied by either General Electric or Westinghouse. Over half of turbine outages are of extremely short duration, 10 hours or less, and are generally caused by seal leakage or turbine control malfunction. Mid-range outages of up to 200 hours are usually caused by valve or lubrication failure, or vibration correction. Longer durations involve time-consuming repair or replacement of turbine blades or rotors.

Westinghouse turbines have a failure rate roughly three times that of GE's, and are almost solely responsible for blading failures which cause long-duration outages.* GE turbines, however, have not been installed on Westinghouse reactors and have been installed only recently on Babcock & Wilcox and Combustion Engineering reactors, so that most of the experience of GE turbines has been accumulated on BWR nuclear steam supply systems rather than PWR's.

2. *Steam Generators* (1.9% capacity factor loss): This item is the largest single contributor to forced outage time for nuclear units. Only PWR's have steam generators, groupings of thousands of tubes that transfer heat from pressurized water in the reactor core to a separate loop containing water which boils to make steam.** They have about a 3% loss impact on PWR's through forced outages. They create a high forced outage time due to maintenance difficulty in their cramped spaces which are, moreover, radioactive on the "primary" side (the side which contains pressurized water from the reactor).

The major cause of steam generator outages is tube leaks resulting from corrosion caused by the normal interaction of tube surfaces with reactor coolant water (containing associated impurities, e.g., fission products from the nuclear reaction) or by caustic corrosion in restricted flow areas, such as tube sheet crevices, where sludge may deposit. EPRI notes that no pattern has been identified to explain the cause or frequency of steam generator tube leaks during the period covered by their analysis. During this period many PWR's were switching their "water chemistry" control to reduce corrosion.*** Data on results of these conversions—from "phosphate" treatment to "all volatile water" treatment—was considered insufficient by EPRI to permit an assessment of the impact of the change.

Prior to the period surveyed by EPRI, tube leaks occurred in several PWR's which were traced to inadequacies in the "explosive bond" method of applying stainless steel cladding to the carbon steel tube sheets of the steam generators. These leaks led to lengthy outages at several PWR's, notably Point Beach 1 and Palisades in 1971-72, which are not reflected in the 1.9% average capacity

*EPRI notes that the failure rate of one turbine manufacturer was three times that of the other in the period studied.[17] The higher failure rate of Westinghouse turbines at nuclear plants is well known in the power industry.[18]

**The exception is the Dresden 1 BWR, which is a "dual-cycle" reactor utilizing steam on both sides of the steam generators. Dresden 1 is considered a pre-commercial reactor and is not in the CEP data base.

***Water chemistry refers to the use of chemical additives to reduce corrosion-causing impurities in the reactor coolant water which impinges upon steam generator tubes from the "primary side" and in the condenser coolant water which interacts with tubes from the "secondary side."

factor loss associated with steam generators in Table 4.2. Explosive bonding has subsequently been discontinued as a means of applying cladding material.*

3. *Condensers* (1.3% capacity factor loss): Condensers are more than a direct source of forced outage time. According to EPRI, they exert indirect impact by (1) leakage which abets corrosion of steam generators by creating impurities on the secondary side of steam generator tubes; (2) partial power reductions while condenser tube leaks are being repaired, which are not reflected in full forced outage statistics; and (3) frequent repairs during "acceptance and pre-operation testing"—pre-commercial operation periods which are not included in the data bases of EPRI or of the present study. EPRI offered no "diagnostic" comments on condensers, largely because of the diversity of condenser designs, materials, and cooling water in operating nuclear plants.

4. *Valves* (1.4% capacity factor loss): Each of the thousands of valves in nuclear plants has a relatively low failure rate, but their number and occasional propensity to cause long-duration outages make valves a significant source of plant outages, according to EPRI. Steam service valves have the highest failure rates, particularly "main steam isolation valves" (three feet in diameter, they keep steam in the reactor core in case of a loss-of-coolant accident); "safety/relief valves" (intended to provide venting in the event of overpressure); and valves performing control and regulation functions. However, EPRI notes that, ". . . the variety of 'valve' problems experienced . . . was very broad. This latter consideration demands a very difficult further study simply to understand what is taking place and to identify the specific problem causes on an individual basis."[19]

5. *Pumps* (0.6% capacity factor loss): As with valves, pumps have relatively low failure rates but nevertheless contribute to capacity factor loss because of the number of pumps per plant. However, the major source of pump outages is identified by EPRI. BWR recirculation pumps and PWR reactor coolant pumps, which continuously drive primary coolant water through the reactor core, are responsible for most outage time attributable to pumps. Seal or drive motor failures cause most outages.

6. *NSSS* (4.2% capacity factor loss): Problems with the nuclear steam supply system, or reactor core, have generally been responsible for significant losses in capacity factors. These losses typically include 0.7% for forced outages of core equipment, and 3.5% for partial power reductions from fuel failures. However, the 4.2% total does not include capacity factor lost when refueling is rescheduled to perform in-core repairs. Such losses are reflected in refueling outage statistics.

The 0.7% loss for NSSS forced outages is primarily due to failure of control rod drive mechanisms, in both BWR's and PWR's; to defects in reactor cores and reactor pressure vessels (leading to excessive vibration), which have been more common in BWR's; and to feedwater sparger failures and pipe cracking in steam-by-pass lines. Aside from the last item, which has occurred in a number of BWR's, EPRI notes that NSSS failures tend to be variable in nature and not readily characterized.[20]

*During the 30-month EPRI survey period Palisades experienced forced outages totalling 17 months in duration. Though related to steam generator problems, they did not result directly from the original explosive bonding of tube cladding.

Partial power reductions due to fuel failures appear to account for 3½% in reduced capacity factor at typical nuclear plants.[21] Fuel failures include structural damage to the nuclear fuel pellets or their structural cladding—such as fuel densification or flattening, or cladding structural changes induced by corrosion or radiation. Power reductions are often required during fuel failures to hold radiation release rates (either off-site or in-plant) within satisfactory limits, or to avoid high "peak" temperatures in the core which are unacceptable under the NRC loss-of-coolant criteria.

Such de-ratings appear to be more common at BWR's than PWR's, although comprehensive data is not readily available. A number of BWR's have operated within restrictive limits governing the rate of change of reactor power level. This avoids build-up of local stress on the fuel pellets and fuel rod cladding that could lead to cladding "leakage" (passage of radioactive fission products through the cladding into the reactor coolant water). For example, Quad Cities 1 was until recently limited to power level changes as low as 3 Mw/hr above 400 Mw load, which cost an estimated 37 full power days of operation in 1974. This is equivalent to an average annual 10% capacity factor reduction.[22] The original Quad Cities 7 x 7 fuel matrix, in which each fuel "assembly" consists of fuel rods in a 7 x 7 square array, was replaced by an 8 x 8 configuration during refueling in early 1976 in an effort to reduce such stresses.

Most PWR's have more finely tuned control rod mechanisms and use a boron "chemical shim" to supplement the use of control rods in controlling the neutron flux to change power levels. Thus, PWR's typically have greater load maneuvering capability than BWR's.

7. *Non-classified Outages* (5.2% capacity factor loss): These account for 3.1% of the 10.5% average capacity factor reduction attributable to full forced outages. They include generic events such as steam-flow mismatch in PWR steam generators and instrumentation and control errors, and specific outages such as the Browns Ferry cable tray fire.

In addition, events classified as "All other and unaccounted for" render about a 2.1% capacity factor loss from partial power reductions, based on EPRI data.* Thus, non-classified events account for at least 5.2% in capacity factor lost at typical nuclear units. This is from a total of 40.7% in average lost capacity factor in the CEP data base, 17.5% of which is due to equipment problems.

Section 4.4: Regulatory-Related Outages

The EPRI tabulation of outage causes indicates that "regulatory issues" have contributed 4% to average nuclear capacity factor reductions. EPRI attributes 3% to partial power reductions and 1% to regulatory-induced full outages.

Regulatory-related outages or de-ratings, as defined by EPRI, refer to NRC-mandated inspections, such as those of BWR's in early 1975 to check for

*Other events classified broadly as "Equipment failure related" account for another 1.4%. They presumably include partial failures of main equipment such as condensers, steam generators, and turbines.

pipe cracks and leaks in by-pass steam lines and related equipment;* to outages or de-ratings to avoid exceeding the "technical specifications" contained in each nuclear plant license, such as de-rating to avoid excess radiation releases; and to NRC-mandated regulatory training.

EPRI also notes that regulatory requirements could be construed to account for 10%-15% in reduced capacity factor if one views NRC regulations as overly conservative.[24] For example, Trident Associates cites chronic problems with "sway arresters" (shock absorbers designed to mitigate earthquake impact), which have required frequent shutdowns for repair at Oyster Creek. An NRC regulation prohibits operation for more than seven days after detection of a failed sway arrester on a safety system. Trident notes that this restriction could be considered overly conservative in view of the alleged quiet seismic history within the plant vicinity. This regulation has cost an estimated 1100 hours of forced outage time at Oyster Creek, or the equivalent of a 2% annual capacity factor loss over the plant's six-year history.[25] Such outages are generally assigned by EPRI to scheduled or forced outage categories rather than to regulatory outages.

Section 4.5: Load-following and Time to Come to Full Power

These are actually two separate categories of partial power reductions. "Load-following" refers to discretionary operation at less than full power because of inadequate system demand for electricity. "Time to Come to Full Power" refers to operation at less than full power as the reactor is brought from start-up to full load. According to EPRI figures, each category has caused the typical nuclear plant to operate at 1% less than full capacity during hours when the unit was available to operate. Since EPRI estimates average nuclear plant availability during its survey period to be 71%, Load-following and Time to Come to Full Power have each been responsible for about 0.7% average capacity factor reduction.**

The estimated 0.7% impact of Load-following is important in that it establishes that nuclear units have virtually always been fully base-loaded when mechanical and regulatory conditions permitted. The capacity factors of nuclear units to date are thus an accurate measure of their "achievable capacity factor" or "reliability" and can be used as a starting point for future projection.***

*Roughly 60% of these inspections were actually performed during refueling outages and were not classified as forced outages. Cracks were detected in several by-pass lines which were subsequently removed and replaced.[23]

**The EPRI estimate actually translates to a range of roughly 0.5% to 1.0% due to the statistical significance of the 1% figure for the estimated reduction in output during available hours.

***Two other surveys confirm that nuclear units have rarely operated in load-following fashion. The S.M. Stoller Corporation estimated that the average capacity factor reduction for nuclear plants due to load-following was 0.5% or less during 1970-1974.[26] Similarly, the Federal Energy Administration determined from an evaluation by the NRC of U.S. reactors in a sample month (August 1974) that the average capacity factor intentionally "foregone" by nuclear units was 0.6%.[27]

REFERENCES

1. M.E. Lapides, E. Zebroski, "Use of Nuclear Plant Operating Experience to Guide Productivity Improvement Programs," Electric Power Research Institute, EPRI SR-26-R, November 1975.
2. D.G. Bridenbaugh, R.L. Turner, G.B. Lloyd, "Maintenance and In-Service Inspection Experience at Large Nuclear Power Plants," in *Peaceful Uses of Atomic Energy*, Proceedings of the 1974 IAEA Conference, 1975, p. 322.
3. Nuclear Regulatory Commission, "Operating Units Status Report," (The "Gray Book"), April 1975, p. 2-25, and telephone communication with Frank Palmer, Commonwealth Edison, Manager, Division of Nuclear Stations, July 1, 1976.)
4. Lapides & Zebroski, *op. cit.*, p. B-29.
5. Trident Associates, "Power Plant Productivity—An Assessment of Outages at 2 Coal-Fired and 2 Nuclear Plants," November 17, 1975, p. J-4.
6. Refueling shutdown and startup dates are from the monthly NRC Gray Books.
7. H.L. Ottoson, W.R. Gould, "Operating Experience with San Onofre Nuclear Generating Station," p. 15, in *Peaceful Uses of Atomic Energy*, 1975, *op. cit.*
8. Bridenbaugh et. al., *op. cit.*, p. 321.
9. Telephone communication with Francis Wiedenmann, Nuclear Information Division Head, Boston Edison, June 30, 1976.
10. Telephone communication with Sol Burstein, Executive Vice President, Wisconsin Electric Power, June 28, 1976.
11. Bridenbaugh et al., *op. cit.*, p. 322.
12. Lapides & Zebroski, *op. cit.*, p. B-29.
13. R. Billingham, "Fifteen Years in Nuclear Power," p. 182, in *Peaceful Uses of Atomic Energy*, 1975, *op. cit.*
14. Lapides & Zebroski, *op. cit.*, p. 16.
15. Lapides & Zebroski, *Ibid.*, p. B-25.
16. *Ibid.*, p. 35.
17. *Ibid.*, p. B-8.
18. See, for example, S. M. Stoller Corporation, "Performance Characteristics of Nuclear Power Plants in the United States," March 6, 1976, p. 19.
19. Lawrence E. Minnick, Director of Nuclear Engineering and Operations, EPRI, "Operating Experience With Light Water Reactors in the USA," April 1975, pp. 14-15.
20. Lapides & Zebroski, *op. cit.*, p. B-22.
21. *Ibid.*, p. 16.
22. Nuclear Services Corporation, "Productivity Improvement Action Plan," report prepared for FEA, November 28, 1975, p. II-3.
23. Telephone communication with Mel Lapides, EPRI, June 30, 1976.
24. Lapides & Zebroski, *op. cit.*, p. 37.
25. Trident Associates, *op. cit.*, 1975, pp. K-4 and K-5.
26. S.M. Stoller Corporation, "Survey of Operating Statistics of Light Water Reactors in the United States," prepared for Long Island Lighting Company, April 11, 1975, Tables 4 and 5, Line J.
27. Federal Energy Administration, "A Report on Improving the Productivity of Electric Power Plants," March 1975, p. 29.

5: The Prognosis for Nuclear Plant Capacity Factors

Section 5.1: Key Findings: Future Performance of Nuclear Units

1. Depending upon design concepts employed at large nuclear units, their lower average operating reliability as compared to small units results from (i) more components of fixed size; (ii) more sets of components of fixed size in more demanding service conditions; (iii) more large components inherently more difficult to assemble and repair.

2. The tendency of commercial PWR's to operate at increasing capacity factors with increasing unit age is well established, but is based only on reactor operation through age eight. Pre-commercial 200-Mw reactors have had sharply reduced capacity factors after age 10, due to wear rather than regulatory restrictions. However, this performance drop cannot be applied with confidence to predictions for current reactors.

3. Occupational exposure to radioactivity in nuclear plants is an unquantified but significant cause of lengthy outage durations. Exposure increases, unless minimized by more shielding or effective design, may result in further capacity factor reductions. There is the danger of manpower shortages in some circumstances, or costly plant decontamination procedures to avoid them.

4. Nuclear plants may improve capacity factors as results of operating experience are fed back into the design, construction, and operation of new plants. Increased economic incentives should result in capacity factor increases via higher construction and maintenance expenditures. Unit duplication has increased PWR capacity factors, and reactor design standardization may also lead to higher capacity factors for new reactors.

5. Potential increases in nuclear capacity factors may be limited by reactors' technical complexity, missing clues to causes of outages, possible increases in regulations on construction and operation, and levels of in-plant occupational exposure.

Section 5.2: Unit Size and Capacity Factor

Each 100 Mw increment in nuclear unit capacity is associated with an average capacity factor decline of about 3.3% for BWR's and PWR's.* Separate equations were used for the two reactor types, which also tested for impact of other potential explanatory variables: unit age and year of commercial installation.

There is little doubt that unit size has exerted a depressing effect upon nuclear plant capacity factors. While unit size has not been statistically related to capacity factor in separate trend equations for first-year PWR's or first- or second-year BWR's, capacity factor has had a strong inverse correlation with unit size for second-year PWR's and for both reactor types beyond their second year of operation.**

Discussion of inverse relation of size to capacity factor has appeared in none of the published analyses of nuclear plant performance by industry spokespersons or federal agencies, including reports by the Electric Power Research Institute and the three consultant studies performed for the Federal Energy Administration.*** One capacity factor survey, widely circulated in the nuclear industry, claims, "A plot of reactor capacity factor vs. size does not seem to indicate a relationship between these two parameters."[2] The article, however, contains no analysis to support the statement.

The failure of industry and government-contracted technical evaluations to address the capacity factor/unit size question results in a paucity of engineering analysis to help explain its impact. Problems of large units are explained by three broad hypotheses, coincident with design approach.

1. *Components in Parallel*—To some extent, large nuclear units tend to use the same overall design approach and employ the same types of components as smaller units, achieving higher power ratings by constructing more componenets in parallel. For example, large PWR's tend to have four steam generator "loops,"**** compared to two or three loops in smaller plants. Similarly, large units generally have more fuel rod assemblies to enable more nuclear fissions to occur, producing more heat, although some uprating has been accomplished by increases in power density.

The intention is to build upon successful experience with design and equipment in existing units. However, to the extent that there is no improvement in component reliability, numerical increase leads to more frequent failures per plant. The failure rate "per megawatt" may be constant, but mechanical and regulatory constraints frequently dictate that nuclear units shut down when certain components fail, rather than operate at reduced power. Forced outages will then be more frequent at large units.

2. *More Demanding Service*—Large units are not necessarily designed with multiple components of fixed capabilities, since this would cut down on

*See Sections 3.3 and 3.4. If seasonally adjusted nuclear capacity factor data for the first half of 1976 is added to the data base, the capacity factor decline per 100 Mw of capacity becomes 4.0% for PWR's, and 5.3% for BWR's (see Appendix D).

**For PWR's, the coefficient of unit size is minus 3.0 percentage points per 100 Mw for AGE = 2, and minus 7.9 per 100 Mw for Age = 3+. For BWR's, the unit size coefficient is minus 9.6 for Age = 3+. (see Appendix E).

***In contrast, papers published in 1975-76 by David Comey of Business and Professional People for the Public Interest and by CEP have consistently pointed out the negative relationship between nuclear unit size and capacity factor.[1]

****Tubing systems in which heat in pressurized reactor water is transferred to boiling water in a separate cycle.

potential economies of scale in construction. Some large unit equipment is comparable in size and capability to that in small units, but may be forced into more demanding service.* Increased failure rates may result from greater duty requirements on components such as piping, valves, and pumps. This may especially enhance "stress-corrosion cracking," a phenomenon in which stainless steel may leak or crack due to the application of high stresses in a corrosive environment. Or it may result in more severe vibration loading due to flow-induced forces and resultant failures. Material cracking has been a significant cause of equipment failure in reactors, especially in BWR circulating systems and PWR steam generators.[3]

3. *Larger Components*—Components specifically sized for large unit function may introduce reliability problems. Beyond certain size limits, some components cannot be fully assembled by the manufacturer. Factory or transportation constraints require site assembly. Such "field assembly" is often performed by less expert personnel in a less controlled work environment and may lead to more common operating failures.

Moreover, difficulty in removal, problem diagnosis, and equipment replacement varies directly with physical size. For example, startup of Philadelphia Electric's 1065-Mw Peach Bottom 2 was delayed by logistical complexity in removing a faulty generator stator to permit a part exchange with unit 3, then under construction. Each stator weighed 450 tons, and the utility had to undertake studies just to find a transport method that properly distributed weight over the building foundation.[4] In general, large pieces of equipment tend to be more difficult to repair and maintain and require longer outages when necessary repairs are performed.

While each design approach is subject to pitfalls which may broadly explain much of the reduced capacity factors of large nuclear units, there is yet to be a comprehensive study of specific outage problems at large units.

Within the power industry it is believed that increased experience and standardization will bring capacity factors up to the 70%-80% range (see Section 5.5). Given the CEP correlation equations' inverse link between unit size and capacity factor, it falls to the industry's power engineers to empiracally illustrate this contention. Unfortunately, the debate over unit size and capacity factor has yet to be joined publicly and cannot be fully addressed in this study.

Section 5.3: Unit Age and Capacity Factor

Also unresolved is the issue of unit age effect upon plant capacity factor. Unit age has not been correlated with capacity factor for BWR's. But the PWR curve that fits closest to historical data (Equation 3.1) indicates significant improvement with increased age (see Sections 3.3 and 3.4). Extrapolating Equation 3.1 to age 10 indicates an average 10 percentage point increase in PWR capacity factor. Half of the increase occurs by age two and four-fifths by age four. The average "levelized" PWR capacity factor over the initial 10 years of commercial operation would be seven percentage points higher than the first-year. The average PWR has clearly profited from a "learning"

*For instance, by taking more water flow from the same pumps through the same number of loops, rather than adding loops.

experience with plant operation, maintenance, and repair, unlike BWR's which have not demonstrated a learning pattern.

Unfortunately, age analysis is impeded by the absence of "old" reactors in commercial sizes. The CEP data base has two PWR's with eight years of commercial operation, two BWR's with six years, and only three other units with five years of commercial experience. And so there is no basis for projecting capacity factors beyond age eight and only a narrow one for ages five to six.

Outside the CEP data base, three U.S. reactors of near commercial size have operated for over 10 years: 200-Mw Dresden 1 of Commonwealth Edison (installed in 1960); 175-Mw Yankee Rowe of New England Electric (1961); and 265-Mw Indian Point 1 of Consolidated Edison (1962).*

These units are excluded due to their smaller size, earlier construction, and different design from later models.** Caution should consequently be used in extrapolating from these older units. Still, the considerable decline in capacity factor for these units following their tenth calendar year of operation, as shown in Table 5.1, merits further analysis.

Average capacity factor of the three units following age ten has been 18.1 percentage points less than during their "mature" period (ages three to ten). Capacity factor declines have been about equal (17%-23%) for each plant. The declines are due primarily to equipment failure at each plant, exacerbated at Indian Point 1 by radiation buildup in the primary system.

Indian Point was shut down all of 1973 for major maintenance. Work included replacement of connecting pins on control rod drives and repair of leaking downcomers (tubes which return condensed steam to the reactor) in steam generators. Rewelding the downcomers took seven months. An intense radiation field developed from corrosion product deposits in the steam generators which limited allowed repair time to about 15 minutes per person at the work site, under AEC occupational exposure regulations. As a result, about 1500 different workers were employed to perform repairs. This led to major inefficiencies and extended outage time beyond that required in a less radioactive environment. Aging helped create wear which led to the Indian Point 1 steam generator leaks and helped prolong outage duration (see Section 5.4).

Yankee Rowe's reduced capacity factor after its 10th year resulted from two problems with reactor core internals. In 1972-73 failed bolts on the control rod shroud package, a device which protects reactor control rods from the influence of circulating water, had to be replaced. The work was performed underwater to limit radioactive exposure, which added to the six month outage. The 1974 refueling outage took nearly four months due to replacement of an in-core instrumentation system. The failures appear normal with age, but their duration was not a function of unit age, according to plant personnel. Primary system

*Indian Point 1 has been shut down since October 31, 1974 due to inability to meet NRC criteria for emergency core cooling.

**For example, Dresden 1, with a GE reactor, is the only U.S. "dual cycle" BWR (with a direct-boiling cycle and a steam generator cycle). Indian Point 1, a Babcock & Wilcox PWR, has an oil-fired superheater which accounts for one-third of its total electrical capability. Yankee Rowe, although similar to later, larger Westinghouse PWR's such as Connecticut Yankee, operated for most of the 1960's with a far simpler backup reactor cooling system than its present version which was installed to meet AEC criteria.

TABLE 5.1: CAPACITY FACTORS OF THREE OLD NUCLEAR UNITS, BY UNIT AGE, THROUGH 1975

Unit	Capacity Factor Ages 1-2	Capacity Factor Ages 3-10	Capacity Factor Ages 11-15
Dresden 1 (BWR)	49.1%	58.5%	37.8%
Yankee Rowe (PWR)	53.2%	79.3%	61.9%
Indian Point 1(PWR)	30.2%	52.0%	28.8%
3-unit Average	44.2%	63.3%	42.8%

radiation levels in repair areas have not increased significantly over the lifetime of Yankee Rowe and so were not a factor in the repair time.[5]

General wear was the major cause of low capacity factor at Dresden 1 in years 11 through 15. Frequent fuel seepage and chronic leaks on a steam generator commonly led to forced outages and partial power reductions to reduce off-gas radioactive releases. Diagnosis was time-consuming as it was difficult to know whether faulty core behavior or intrinsically defective fuel was responsible for fuel leakage. It was also hard to obtain parts to repair a steam generator built over a decade earlier. Commonwealth Edison holds that radiation buildup in the steam generators did not impede repairs. "We had to use more people to do the job," says the company, "but we didn't have to extend the duration of repairs appreciably."[6]

Clearly, equipment failure was the primary agent for the old units' average 18% capacity factor decline after age 10. Regulatory outages were insignificant.* Despite differences in designs from current reactors, the extent of forced outages at all three units may herald capacity factor problems for subsequent nuclear units as they reach comparable ages.**

Section 5.4: Radioactivity and Capacity Factor

Radioactive isotopes in plant components have increased down-time at Dresden 1, Yankee Rowe, and Indian Point 1 in recent years (see Section 5.3). Repair time for steam generators and reactor internals has increased with worker exposure procedures (such as at Yankee Rowe). In-plant radiation often acts as a significant constraint on nuclear plant capacity performance." [The] impact [of radiation] on time to maintain and repair is certainly significant," notes EPRI, although impact "is very difficult to determine in any quantitative way."[7]

Radioactive isotopes, or "radionuclides," are continuously produced in nuclear plants in two major ways. First, the fission chain reaction splits uranium atoms into radioactive "fission products." Some fission products may leak through the cladding of fuel rods and be deposited in the "primary coolant" (circulating water in which fuel rods are submerged). Second, radionuclides called "activation products" result when impurities in primary

*As noted, this figure excludes Indian Point 1 since its October 31, 1974, shutdown.

**Coal-fired units in the 100-399 Mw class have shown only about a 1% per year fall-off in capacity factor after age 10 (see Section 7.5).

coolant water pass through the neutron flux in the reactor core. Impurities derive from corrosion in the reactor structure created by dissolved oxygen in primary coolant water or friction erosion by water at high temperatures, pressures, and flow rates.

Nuclear plants have complex removal systems to rid primary coolant of radionuclides by transfer to treatment equipment. Some invariably escape, however, and are deposited onto plant components. This is particularly so on the "primary" or reactor side of PWR steam generator tubes, on the inner surfaces of recirculating piping systems (both PWR and BWR), and on BWR steam line surfaces. Radiation fields are always present in nuclear plants, inside the reactor vessel, and in primary system components. Equipment leaks such as in steam generator tubes often transport radionuclides to the secondary system as well.

NRC regulations limit nuclear power plant workers' exposure to three "rem" ("Roentgen Equivalent Man") per quarter. Annual 12 rem exposure is permitted to a cumulative dose equivalent of five rem per year. Here, exposure is further limited to an annual rate of five rem.* These regulations dominate the logistics of nuclear plant deployment of highly skilled personnel, particularly in maintenance operations which account for most radiation exposure.

Indian Point 1 attempts at in-place repair of defective steam-generator welds had to be abandoned due to the cramped work location and high radiation fields (up to 15 rem/hour) resulting from deposits of activated corrosion products on tubing (see Section 5.3). The radiation field restricted the average worker's repair time to 20 minutes, including five minutes to crawl into and out of the repair area through radioactive spaces. Severe space limits hampered efforts to reduce radiation with lead shielding and decontamination with caustic chemicals was ruled out for fear of equipment damage. Defective tubing had to be cut out and repaired off site.

Finally, reinsertion could not be done by Con Edison personnel due to exposures they received during the initial repair attempt. Use of less experienced outside tradesmen (which eventually included almost every union welder in New York City and adjacent Westchester County) in a cramped, high-radiation space under severe time pressures, resulted in low productivity. Continuous work force changes led to time-loss confusions such as mistaken removal of two good downcomers and radiographic examination of the wrong welds.[8]

This incident, though extreme, is by no means unique. In 1971 rewelding of steam generator tube sheet cladding in a high radiation field at Carolina Power & Light's Robinson 2 unit involved 255 non-company personnel and took two months.[9] Repair or change of cracked feedwater spargers** at Millstone 1, Pilgrim 1, and several Commonwealth Edison units have required extended outages, in part because of radiation fields in the reactor vessel.

*Maximum permissible cumulative dose limit is (N-18) × 5 rem, where N is the person's age. Thus, a person with no previous exposure may accumulate exposure at the rate of 12 rem per year. At that rate, he would draw down his exposure "bank" at the rate of seven rem/year. Once the "bank" has been exhausted, the dose rate of five rem per year applies.

**These are pipes in the upper portion of the reactor vessel which feed water into the core through small holes.

Indian Point 1 had another major radiation-extended outage in 1970, involving piping near the boundary of the reactor vessel. Lead shielding in corridors and over pipes reduced radiation levels to 0.3 rem/hour from levels of 1-12 rem/hour. Nevertheless, over the course of eight months the job took all 60 of Con Edison's skilled welders, 50 health physicists to monitor exposure, and about 600 additional personnel. It is estimated that a similar effort in a conventional plant would have taken about two weeks and 25 workers.[10]

High-radiation repairs coincide with inherent inefficiency in rapid repair personnel sequencing and so long outages result. Use of protective clothing and shielding such as lead or water can somewhat reduce exposure levels. Remote handling requires expensive equipment and introduces a "hindrance factor" estimated to be 3-10/1, compared to direct manual operation.[11]

In-plant radiation levels are primarily related to fuel performance, primary coolant behavior, and remedial measures to reduce exposure. Efforts to reduce outage time they require are recent and it is too early to judge their impact. According to EPRI "improved" fuel designs with which vendors hope to reduce leakage of fission products from fuel have yet to be fissioned to a high enough "exposure" or "burn-up" for their performance to be statistically significant.[12]

Also embryonic are changes in "water chemistry," the use of chemicals to control impurities in primary and secondary coolant, which has not conclusively reduced corrosion levels in the primary coolant. For example, Southern California Edison added hydrogen peroxide to the coolant system of San Onofre during refueling in 1973. The intent was to dissolve corrosion products such as cobalt isotopes circulating in pipes and filters, thereby restricting personnel access and delaying refueling. No significant change in cobalt concentrations was observed, however.[13]

Occupational exposure can also be reduced, in theory, by designing plants with thicker shielding and more space in key maintenance areas prone to high radiation fields. Similarly, "decontamination" to flush deposited corrosion products from the primary coolant can reduce exposure loads. But the diagnostic work required for most of these design improvements is far from completion.

Decontamination, for example, has yet to be performed on a U.S. nuclear plant greater than 100-Mw capacity. The 200-Mw Dresden 1 unit is tentatively scheduled to undergo a major 6-month decontamination to begin in December 1977.* The cost of Dresden 1 decontamination is estimated by Commonwealth Edison at $25 million.** EPRI, ERDA, and Consolidated Edison are now considering decontamination for Indian Point 1.

Feedback from pilot decontamination projects is several years away. Moreover, EPRI cautions that "decontamination must be tailored to the specific plant" because the required chemicals may have a destructive impact on the oxide coating which builds up on metal surfaces and acts as a protective

*High radiation levels make certain parts of the primary system inaccessible for hands-on work such as ultrasonic weld checks mandated under the in-service inspection requirements, so Dresden 1 has been operating under an NRC waiver for several years.[14]

**About $125 per Kw of plant capacity, although cost of a prototype decontamination may not be indicative of future costs.[15]

layer.[16] Disposal methods must also be found for radioactive waste-bearing solutions.

Not all nuclear plants will require decontamination. Whereas radiation fields at Indian Point 1 increased steadily prior to shutdown, levels at Yankee Rowe which has had far more reliable fuel performance have tailed off in recent years.[17] PWR's often show high radiation levels in steam generators early in plant life, but do not thereafter follow a discernible pattern of in-plant radiation build-up.

BWR's have accumulated radiation fields in key components at a rate of 0.1 rem/hour per "full-power year," according to former G.E. engineer Dale Bridenbaugh. With a "rule-of-thumb" maintenance work inefficiency threshold of 0.6-0.8 rem per hour radiation fields, Bridenbaugh foresees substantial drops in BWR capacity factors between ages 12-15. The alternative, at that point, would be decontamination or substantial plant modification to reduce the radiation field. Both would be costly and involve long shutdowns.[18]

Nuclear plant construction rates have created concern over occupational exposure as an impediment to improved plant capacity factors. Bernard J. Verna, a columnist for *Nuclear News*, wonders "if occasions might arise when plants would not be able to gather sufficient numbers of men (with unused exposure . . .) to complete a job."[19]

Repair incidents at Indian Point 1 led Con Edison to scour its whole system in one case and its entire service territory in another to find enough skilled, unexposed personnel. According to Bridenbaugh, the NRC-mandated inspection of reactor coolant by-pass lines for potential cracks in all 23 BWR's in early 1975 "essentially used up" allowable radiation exposure of most qualified ultrasonic testers in the country. "When you get an industry-wide maintenance or inspection problem like that," says Bridenbaugh, "there just may not be enough people around to take that exposure."*[20]

Section 5.5: The Potential for Improving Nuclear Capacity Factors

Industry spokespersons are virtually unanimous in their optimistic 70%-80% predictions for future nuclear plant capacity performance. Inprovement over today's 60% average capacity factor (50% for plants 800 Mw or larger), is predicated on higher reliability for "mature" plants (commercial units three years or older) and on improved design of plants yet to come on line.

To assess the claims for plant "maturity" we have correlated capacity factor and current unit ages. PWR's have had higher capacity factors with increased age, but no BWR improvements have occurred (see Equations 3.1 and 3.3). However, sparse experience with commercial-size units (450 Mw or larger)

*Generally occupational exposure levels (the product of the intensity of radiation fields in the reactor and the man-hours spent working in these fields) have been stable in PWR's during 1970-75 but have increased steadily in BWR's from initial low levels. Cumulative exposure levels per unit were about the same for the two reactor types at the end of 1975, although 1975 exposure per unit was about twice as high in BWR's due to more maintenance requirements. Source: "Occupational Radiation Exposure at Light Water Cooled Power Reactors, 1969-75," NUREG-76/109, NRC, 1976.

has been had after age five, and none after age eight. Conversely, three older reactors in the 200-Mw class show significant equipment wear after age 10, resulting in lower capacity factors. However, only longer experience can determine if the same will happen to larger reactors of more recent design.*

The nuclear industry pins most of its hopes for higher capacity factors on new unit design being more reliable than that of current units. Nuclear engineers point to many design improvements and quality control techniques to buttress their optimism. Plant operating records can be analyzed to determine what equipment failures most significantly reduce capacity factor. If particular manufacturers are at fault (e.g., Westinghouse turbines have had higher failure rates than G.E. turbines), competitors can be used. If failure rates are high for all manufacturers, then equipment specifications may be upgraded. The resulting increase in capital costs can often be paid back through higher capacity factors. For example, EPRI notes economy has dictated that architect-engineers stipulate a minimum number of purchase specifications for valves. In the future, it may be cost-effective to identify valves with the highest failure rate and raise their specifications.[21]

Moreover, if some outages appear to be intrinsic to the equipment then, ". . . the appropriate productivity orientation is toward predetection and accommodation of failure," according to EPRI. This calls for on-line diagnostic instrumentation, redundant equipment, improved maintenance access, and adequate spare parts inventories.[22] Outage duration may also be reduced by design or maintenance changes to reduce radiation exposure (e.g., permanent and portable shielding combinations in high-maintenance areas).[23]

Furthermore, says the industry, a "learning curve" exists for some equipment. Though steam generators are still a chronic source of PWR forced outages (accounting for about a 3% average capacity factor loss from January 1973 to June 1975), their failure rate is less than the pre-1973 record. Certainly steam generator tube failures that added to the average 23.2% capacity factor at Palisades have not recurred with similar intensity at subsequent plants.** The

*The two oldest commercial reactors in the U.S., the 450-Mw San Onofre and the 575-Mw Connecticut Yankee, will begin their eleventh years of commercial service in 1978. Both units are among the most consistent nuclear units with lifetime capacity factors of 68.8% and 73.3%, respectively, and even higher capacity factors in recent years.

**Many in the nuclear industry would eliminate Palisades from the statistical data base on grounds that its problems were "one-of-a-kind" and resulted from its being the first commercial nuclear unit of the particular manufacturer (Combustion Engineering). A comparison of the PWR regression equations with Palisades (Equations 3.1 and 3.2) and without (see Appendix E) indicates that elimination of Palisades from the data base increases the 10-year levelized capacity factor from the equations by 2.6 percentage points for a 600-Mw PWR, and by 9.0 points for a 1000-Mw PWR by significantly reducing the effect of unit size on capacity factor.

Equations 3.1 and 3.2 are not the only basis for projecting future PWR capacity factors, but it is more realistic to keep Palisades in the data base. Although Palisades' is the most serious, the nuclear industry has been characterized by a number of "one-of-a-kind" problems. The Browns Ferry cable tray fire is a recent example. There are not clear criteria for eliminating any reactor's history, whether good or bad. Elimination of Connecticut Yankee whose 8-year average 73.3% capacity factor puts it at the high extreme of the sample, would reduce projected PWR levelized capacity factors from Equations 3.1 and 3.2 by 3.4 percentage points for 600-Mw units and 5.1 points for 1000-Mw units, or almost the reverse of the impact of dropping Palisades. Moreover, Combustion Engineering was not a newcomer to the nuclear field, having designed reactors for the Navy's nuclear submarine program.

industry argues that similar improvements have been made with control rods, nuclear fuel, and reactor core and pressure vessel integrity.[24] General improvement in capacity factors for new units compared to old has not been observed in the CEP statistical analysis even if unit size and unit age are effectively held constant using regression analysis.

The expectation that a capacity factor learning curve will soon be demonstrated is the major industry defense of its decision to build, almost exclusively, very large nuclear units (1000+ Mw) in the future. For example, Southern California Edison, with fairly high capacity factors at its 450-Mw San Onofre Unit 1 (68.8% during eight years of operation, 74.1% without the first year), has ordered two 1140-Mw units for San Onofre 2 and 3. It anticipates that the large nuclear plants will follow the same capacity factor growth curve as have successively larger oil-fired units it has constructed in the past 30 years. SCE has found that, "there is a capacity factor learning curve when a design/construction team builds identical units for a long enough time that operating experience from the first units of a set is fed back into the design of later units."[25]

SCE and other utilities are not convinced that early large units (800+ Mw) in service or under construction will not improve on their present low capacity factors. "The maturation period for large nuclear plants may be four years," says Jack Moore, Vice-President for Advanced Planning, noting that only two 800+ Mw nuclear plants have operated commercially for four full calendar years.[26]

Records of the four commercial units which have operated for six to eight years, shown in Table 5.2, indicate that older, smaller nuclear units have required only one to two years to iron out their initial problems. However, Moore's thesis cannot be proven or disproven from current experience. Moreover, Moore and many others in the industry regard the early large units as "commercial prototypes" which will guide design and operating improvements in subsequent units.

The general hypothesis that nuclear capacity factors will follow a learning curve is supported by an analysis of "paired" nuclear units in which an identical second unit was commercially installed by the same utility, architect-engineer, and construction manager about a year after completion of the first unit. There are six such PWR pairs.* The capacity factor of the second unit has exceeded that of the first unit in each case, with the increase averaging 6.5 percentage points (see Table 5.3). This contradicts results implied by the positive relationship between PWR unit age and capacity factor.

Table 5.3 suggests that duplicate units' capacity factors are higher by their being devoid of problems which occurred early in the operation of their older companion units. Wisconsin Electric Power Company which operates Point Beach 1 and 2 reports that, "A single event is responsible for the capacity factor

*Actually, there are seven "duplicate" units since Oconee 2 and 3 are both duplicates of Oconee 1. For BWR's there are only two initial/duplicate pairs—Quad Cities 1 and 2 and Peach Bottom 2 and 3. Dresden 2 and 3 do not qualify since Dresden 3 was begun after Dresden 2, but entered commercial service earlier. Although BWR duplicate unit experience is too limited to test for trends, the Peach Bottom stator "exchange" described in Section 5.2 demonstrably reduced a pre-commercialization delay at Unit 2 without delaying start-up of Unit 3.

TABLE 5.3: CAPACITY FACTORS OF PAIRED PWR UNITS

Units	Commercial Date	Age on 12-31-75	Capacity Factors 1st Year	2nd Year	3rd Year	Average	Capacity Factor Improvement
Zion 1	12-31-73	2	37.8%	53.4%	-	45.6%	
Zion 2	9-17-74	1	52.5%	-	-	52.5%	+6.9%
Oconee 1	7-15-73	2	51.5%	68.1%	-	59.8%	
Oconee 2	7-11-74	1	64.0%	-	-	64.0%	+4.2%
Oconee 3	12-16-74	1	64.9%	-	-	64.9%	+5.1%
Prairie Island 1	12-5-73	2	30.9%	79.6%	-	55.2%	
Prairie Island 2	12-21-74	1	68.4%	-	-	68.4%	+13.2%
Turkey Point 3	12-14-72	3	50.7%	55.5%	67.0%	57.7%	
Turkey Point 4	9-7-73	2	65.8%	61.1%	-	63.5%	+5.8%
Surry 1	12-22-72	3	48.3%	46.0%	54.3%	49.5%	
Surry 2	5-1-73	2	36.5%	70.1%		53.3%	+3.8%
Point Beach 1†	12-21-70	5	75.2%	67.6%	63.0%	69.0%	
Point Beach 2	9-30-72	3	68.7%	73.0%	85.9%	75.9%	+6.9%
6 Initial Units		17				58.1%	
7 Duplicate Units		11				64.6%	+6.5%

†Capacity factors for Point Beach 1 in years four and five, not given above, were 72.2% and 67.1%.

TABLE 5.2: CAPACITY FACTORS OF FOUR OLDEST COMMERCIAL NUCLEAR PLANTS

Unit	Type	Capa-city	# Years	1st Year	2nd Year	3rd Year	After 1st year
San Onofre	PWR	450	8	32%	66%	78%	74%
Connecticut Yankee	PWR	575	8	59%	72%	70%	75%
Oyster Creek	BWR	650	6	61%	67%	66%	65%
Nine Mile Point 1	BWR	610	6	32%	55%	61%	60%
4-unit Average				46%	65%	69%	69%

difference between the two units. We lost five months for steam generator repairs on Point Beach 1 (in 1971-72), but we were able to make the same repairs on Unit 2 in six to eight weeks, in a non-radiological environment, before it went into service.''[27]

Similarly, Mechanics Research , Inc. reported in its power plant reliability study for FEA that, ''A number of problems experienced on [Turkey Point] Unit 3 were corrected on Unit 4 before any loss of productivity was experienced . . . . In addition . . . personnel experience at Unit 3 has provided considerable training for related work on Unit 4. This has resulted in a higher level of proficiency of Turkey Point personnel in a shorter period of time than would normally have been acquired on a single unit.''[28]

Table 5.3 and the sources cited suggest that duplicate units will reach maturity faster than their companion units, but will not necessarily maintain their initial capacity factor advantage. Furthermore, the data base of duplicate unit operating experience is extremely small.*

*The duplicate unit category is statistically related to capacity factor for PWR's in the CEP data base:

(Equation 5.1)

$$\text{Capacity Factor} = 95.45 - \underset{(1.117)}{3.502}\,(\text{Mw}/100) - \underset{(2.39)}{5.43}\,\text{AGED} + \underset{(5.10)}{10.00} \times \text{DUPE},$$

where ''DUPE'' = 1 for duplicate units and zero otherwise, and AGED = 1/Log (1 + AGE), with an Adjusted R^2 of 0.25, N = 66.

On a 10-year levelized basis, the equation becomes:

(Equation 5.2)

$$\text{Capacity Factor} = 86.28 - 3.502\,(\text{Mw}/100) + 10.00 \times \text{DUPE}$$

From this equation, the capacity factor of a duplicate unit would be expected to exceed that of its older twin by 10 percentage points. For an 800-Mw PWR, Equation 5.2 would predict a levelized capacity factor 1.2 percentage points less than Equation 3.2 for a non-duplicate unit, but 8.8 points

However, experience to date does suggest higher capacity factors for newer units will coincide with accelerated design standardization.

Plant standardization, in fact, is a major basis of the optimism in the power industry that nuclear capacity factors can significantly improve. The emergence of standard reactors, anticipated in the mid- to late 1980's, is expected to reduce reactor-related outages as compared to the current use of customized designs.* Development of optimal designs, selection of proven components, fewer construction, operation, and maintenance errors by increased personnel familiarity with equipment, and off-the-shelf availability of spare parts, are generally expected to result in higher capacity factors for standardized reactors. The "balance-of-plant"—integration of structural, electrical, and piping systems assembled by the architect-engineer—is less amenable to standardization than the reactor. But it is anticipated that the general design and construction of the non-nuclear portion of the plant and its interface with the reactor will be improved by reactor standardization.

Standardization does have detractors. Nuclear critic Jim Harding, an energy specialist with the California Energy Commission, argues that, "Standardized design can result in standardized failures," insofar as many of the causes of current equipment failures are not fully understood. Unit duplication, too, can extend the impact of design defects, such as the extended outages at Browns Ferry 1 and 2 resulting from the cable tray fire below the common control room. Moreover, "certain types of reliability may be more important than capacity factor," says Harding. He cites the nationwide shutdown of BWR's in early 1975 to inspect for possible cracks in steam lines as a potential forerunner of "generic design deficiency" outages which could impair reliability of electric service in the future.[29]

Harding's skepticism of the benefits of nuclear plant standardization is a minority view, however. Most industry observers anticipate that the higher capacity factors observed for the small sample of duplicate PWR's will extend and perhaps increase as units are built from standard designs.

---

higher for a duplicate unit.

Equations 5.1 and 5.2 were not used in statistical analysis for PWR's because (1) sample space of duplicate units is small: seven units, 11 unit-years (2) duplicate units will probably exhibit less maturation than was shown by the average PWR, which is reflected in the "AGED" term in Equation 5.1. The equation effectively "gives credit" to duplicate units for both maturation and being a duplicate unit.

If seasonally adjusted capacity factor data for the first half of 1976 is added to the data base, duplicate units are predicted by regression equation D4 to have capacity factors only 6.9% higher than non-duplicate units and only 5.6% higher than predicted for all units, based on an 800-Mw unit size. Moreover, the 6.9% coefficient in Equation D4 is significant only to the 93% level, or less than the 95% level used as cut-off point in this study (see Appendix E).

*Each of the four U.S. reactor vendors has recently submitted several designs for standardized "nuclear islands"—the reactor vessel, steam generators (for PWR's), and associated pumps, pipes, and controls—to the NRC for generic approval in future licensing hearings.

Duplication, standardization,* and learning from information feedback applied to future design and operation are strong grounds for projecting increases in nuclear capacity factors over the current statistical correlation equations (see Chapter 3), particularly for units over 800 Mw. But the capacity performance of nuclear units to date constitutes a specific record, while the magnitude of potential improvements is speculation rather than forecast. The record suggests there are also strong grounds for anticipating that capacity factor improvements will be of limited magnitude.

First, the complexity of the physical environment at nuclear plants creates logistical problems which frequently extend outage time. For example, refueling is a "significant logistical exercise" according to G.E. It means "taking a work force which may peak at several times the normal station work complement and attempting to place them at work with a minimum of lost time."[30] G.E. emphasizes the need for skillful planning, but breakdowns follow no maintenance schedules.

At Boston Edison's Pilgrim 1 plant, replacement of jet pump riser bolts and feedwater sparger added a month to the 4-month refueling and maintenance outage (see Section 4.2). These failures could not be anticipated and were not even discovered until the reactor was opened for refueling. Furthermore, "space problems" and "skilled labor problems" prevented the tasks from being done concurrently. "One job gets in the way of the next job, because it's part of a system that ties into the next system," explained Boston Edison, "and you have to order parts from the manufacturer, which causes built-in delays."[31] It is hard to envision a systems approach to speed such repairs, though standardized plants would reduce the spare parts problem.**

Second, although a learning curve has been observed on some types of nuclear plant equipment, many of the reasons for forced outages are not yet well understood. Of the seven forced outage cause categories analyzed by EPRI (see Section 4.3), four which account for over half of full forced outage time—condensers, valves, nuclear steam supply system, and "nonclassified"—were described as variable in type and not easily characterized. In addition, modifications in water chemistry (which affects steam generator performance) and fuel design at a number of operating plants are too

Duplication involves identical units at a site, with identical reactors and identical balance-of-plant. To date, the first units of a pair have not operated at higher capacity factors than conventional single units, but the second units have done so (see Table 5.3 and Equation 5.1). The second unit benefits from construction and operation experience on the first unit.

Standardization involves identical reactors at various sites, with the balance-of-plant tailored to the reactor and the site. All standardized units would presumably benefit from the standard design, equipment specifications and spare parts supply, with benefits increasing for later units. Other benefits, from improved construction, operation and maintenance, would require some stability of personnel from one job to the next.

A good discussion of unit duplication and standardization is G.F. Cole, "Replication: One Way to Cost and Schedule Savings," *Power Engineering*, July 1976, pp. 44-49.

**Much maintenance and repair, such as refueling and the annual turbine inspection, is performed concurrently. To the extent that outages have occurred concurrently, the operating reliability problems of nuclear units are understated by capacity factor statistics since the time required for minor repairs is "masked" by other concurrent repair work. Concurrent repair of two or more pieces of equipment is not uncommon, such as the 3-month repair of motor stators on control rod drives at Rancho Seco in 1975 during a longer turbine repair.

recent to be conclusively asssessed. "Until you understand water chemistry, you don't want to design a standard reactor," advises EPRI.[32]

Third, the web of regulations within which nuclear plants must be designed and operated will grow. The number of industry and federally imposed codes and standards applicable to nuclear power plant design and construction has catapulted from near 100 in 1970 to 1600 in 1975.[33] Many have required major changes in plant design. Increases in reactor size, which led to imposition of some standards, have effectively stopped with AEC/NRC promulgation of a 3800-thermal megawatt power limitation for reactors—equivalent to about 1300 Mw (electrical). Moreover, as part of its push for more standardization, the NRC has committed itself to minimizing changes in future design requirements. However, there is no guarantee that new standards will not be promulgated to require further design changes—especially if public concern over nuclear safety remains high, and as new reactor problems are indentified.

NRC regulations which apply to operation are also increasing, particularly regarding in-service inspections.[34] Written into the "technical specifications" of each plant's license are hundreds of tests to be conducted regularly, usually annually. Many tests such as work on electrical systems, instrument calibration, and safety components (e.g., relief valves) can be done only during shutdown. The growing number of such inspections will likely increase the number and duration of scheduled outages.

These regulations have evolved out of increased public and regulatory concern over nuclear reactor safety and appear to be generic, not foreign, to the technology. They are manifested directly in increased design and operating requirements and indirectly in time loss to paper-work and communication with the NRC. This routinely uses management time and occasionally adds to outage duration when major repairs must be cleared with the Commission.

Finally, ever-present radioactivity is a "fact of life" that acts to reduce capacity factors. Fuel failures have caused frequent load reductions to restrict in-plant radiation concentrations or off-gas releases. Much maintenance and repair work must be performed in radioactive environments, impeding efficiency and adding to outage time. While measures may be available to reduce in-plant radiation exposure (see Section 5.4), the repair difficulties inside reactors will not be easily reduced. Moreover, BWR's seem to have steady increases in radiation fields with increasing unit age. This will likely increase maintenance time and may ultimately require decontamination procedures. PWR's do not seem to follow this pattern of radiation build-up.

Significant factors, then, limit the extent of future improvements in nuclear capacity performance. With strong hopes for a more reliable future and limits to improvements in the present, the outlook for nuclear capacity factors is, at best, uncertain.

REFERENCES

1. See, for example, David D. Comey, "Future Performance of Large Nuclear Plants," in mid-April 1975 *Not Man Apart,* and the CEP testimony on nuclear economic issues listed in the Bibliography.

2. "Reactor Performance Evaluation," Harold W. Kohn, Ohio Environmental Protection Agency, *Power Engineering,* December 1975.

3. Letter from W. Kerr, Chairman, Advisory Committee on Reactor Safeguards, Nuclear Regulatory Commission, to William A. Anders, Chairman, NRC, February 8, 1975.

4. Bernard J. Verna, "The Peach Bottom Generator Exchange," *Nuclear News,* May 1975, pp. 35-37.

5. Telephone communication with Ed Tarnuzzer, Engineer, Yankee Atomic Electric Company, July 1, 1975.

6. Telephone communication with Frank Palmer, Manager, Division of Nuclear Stations, Commonwealth Edison, July 1, 1976.

7. Lawrence E. Minnick, Director of Nuclear Engineering and Operations, EPRI, "Operating Experience with Light Water Reactors in the USA," April 1975, p. 17.

8. Bernard J. Verna, "Radioactive Maintenance," *Nuclear News,* September 1975, pp. 54-56.

9. Bernard J. Verna, "Radioactive Maintenance, Part 2," *Nuclear News,* November 1975, pp. 52-54.

10. *Ibid.*

11. M.E. Lapides, E. Zebroski, "Use of Nuclear Plant Operating Experience to Guide Productivity Improvement Programs," Electric Power Research Institute, November 1975, EPRI SR-26-R, p. 44.

12. *Ibid.*, p. 45.

13. H.L. Ottoson, W.R. Gould, "Operating Experience with San Onofre Nuclear Generating Station," p. 153, in *Peaceful Uses of Atomic Energy,* 1975.

14. Interview with Dale Bridenbaugh, former Manager, Performance Evaluation and Improvement, Nuclear Energy Systems Division, General Electric Company, May 12, 1976.

15. Telephone communication with Frank Palmer, Manager, Division of Nuclear Stations, Commonwealth Edison, July 1, 1976.

16. Interview with Bill Lavalle, EPRI engineer, May 14, 1976.

17. Interview with Joe Prestele, EPRI engineer, May 14, 1976.

18. Bridenbaugh interview, *op. cit.*

19. Bernard J. Verna, *Nuclear News,* November 1975, *op. cit.*

20. Bridenbaugh interview, *op. cit.*

21. Prestele interview, *op. cit.*

22. M.E. Lapides, E.L. Zebroski, and L. Johnson, "Developing Methods for Establishing Improved Plant Reliability," paper presented at American Power Conference, April 20-22, 1976, p. 10.

23. B.G. Kniazewycz and W.C. McArthur, Carolina Power & Light Co., "Is Your In-Plant Exposure as Low as Practicable," *Power Engineering,* August 1975, pp. 40-43.

24. Lapides & Zebroski, EPRI, *op. cit.*, p. 23.

25. Gene Cramer, Engineer for Advanced Energy Systems, SCE, personal communication, July 22, 1976.

26. Interview, May 18, 1976.

27. Telephone communication with Sol Burstein, Senior Vice-President, Wisconsin Electric Power, June 29, 1976.

28. Mechanics Research, Inc., "Study on the Reliability of Large Coal-Fired and Nuclear Power Plants," December 15, 1975, Volume II, p. 58.
29. Telephone communication, August 4, 1976.
30. D.G. Bridenbaugh, R.L. Turner, G.B. Lloyd, "Maintenance and In-Service Inspection Experience at Large Nuclear Power Plants," in *Peaceful Uses of Atomic Energy*, Proceedings of the 1974 IAEA Conference, 1975, Chapter 4, p. 330.
31. Telephone communication with Francis J. Wiedenmann, Nuclear Information Division Head, Boston Edison, July 1, 1976.
32. Bill Lavalle, interview, *op. cit.*
33. G.F. Cole, "Replication: One Way to Cost and Schedule Savings," *Power Engineering*, July 1976, pp. 44-49.
34. Ottoson and Gould, *op. cit.*

6: Statistical Analysis of Coal Capacity Performance

Section 6.1: Major Statistical Findings on Coal Capacity Performance

Commercial-size U.S. coal-fired power plants have operated at an average capacity factor of 66.9% (13.8% standard deviation). These figures include all 250 coal-fired units of 100+ Mw installed between 1960 and 1972 and operating from 1961 to 1973. The sample totals 1641 unit-years of operation.*

Unlike nuclear units, many of the coal-fired units are not fully "base-loaded" (operated at constant full available capacity). "Capacity performance" is a measure of operating reliability which corrects capacity factor data for "load-following."** It approximates the capacity factor at which a unit would have operated had it been fully base-loaded.*** The test period average for the 250 coal-fired units is 74.8% (standard deviation: 16.1%).

Performance assessment of the wide range of coal-fired units from 100 to 1150 Mw, installed over a 13-year period, and burning many grades of coal, requires statistical analysis. The following are our conclusions:

1. Supercritical coal units employ a design concept developed for large capacity units in the late 1950's, in which the boiler is maintained at exception-

*The CEP coal data base is defined in Appendix A. Also in Appendix A, is each unit's and each utility's lifetime coal capacity factor, as well as annual data for all units 400 Mw+. Capacity factor is defined in Appendix B.

**"Load-following" is *intentional reduction in power output based on low demand for electricity*.

***Capacity performance is defined in Appendix C. Lack of load-following by nuclear units is discussed in Section 4.5.

The coal data base includes less than one unit-year of operation with "scrubbers" (flue gas desulfurization systems or "FGD"), since most scrubbers installed before the end of 1973 were installed on units constructed before 1960. A rough estimate of scrubber effect on capacity performance is 5% reduction (*i.e.*, 3.7 percentage point reduction from the average capacity performance of 74.8%) (see Chapter 8).

ally high pressures to increase thermal efficiency. They have performed at about 4% higher capacity factor and capacity performance levels than subcritical coal units over 400 Mw capacity.* Higher performance of the supercritical units is a bit more pronounced for 600+ Mw units.**

2. Coal-fired capacity performance is inversely related to unit size. Capacity performance declines at an average rate of 1.1 percentage points per 100 Mw capacity increase for supercritical units; at an average rate of 5.6 percentage points per 100 Mw capacity increase for 100-399 Mw subcritical units; and at 2.2 percentage points per 100 Mw increase for 400+ Mw subcritical units. Overall, the average decline is 3.5 percentage points per 100 Mw for all coal units.

3. Coal-fired units generally take a full calendar year of commerical operation to "mature" to highest capacity factor. For 600+ Mw units it takes one to two years. Capacity factors of subcritical units start decreasing around the fifth to seventh year of operation. Supercritical units, and 600+ Mw units of both types, show increased capacity factors through age six. Data beyond that point is limited.

4. Capacity performance is inversely related to sulfur content of the coal burned. It declines at a 1.8% rate for each 1% rise in coal sulfur content. However, capacity performance is positively related to coal ash content for units under 600 Mw at a 1.4% rate for each 1% rise in coal ash content. Capacity performance is generally not related to coal heat or moisture content.

5. Capacity performance of coal-fired units under 600 Mw is significantly less for units installed toward the end of the 1960-1972 sample period. This is not applicable to 600+ Mw units, there being so few prior to the late 1960's. Variation in capacity performance results from different standards for measuring capacity as well as actual performance records.

6. "Duplicate" supercritical coal units of 600+ Mw capacity operate at higher capacity factors than their predecessors.***

7. American Electric Power's supercritical 600+ Mw coal units have had capacity performance averages 9¼% higher than those of comparable units. Conversely, Pennsylvania Electric Company's average 10½% lower than comparable units. Tennessee Valley Authority's 600+ Mw subcritical units have ranked 9½% higher and Commonwealth Edison's 10% lower than units of comparable size, age, and coal quality.****

*Subcritical boilers operate at pressures under 3208 pounds per square inch. Above that "supercritical pressure" the density of steam is equal to that of the feedwater.

**Average capacity performance of supercritical coal units in the CEP data base is 65.3%, or *less* than average capacity performance of subcritical units (76.7%). This is statistically due to larger average size of supercritical units (587 Mw over 272 Mw for subcritical). Large supercritical units have had higher capacity factors than subcritical units of comparable Mw capacity and installation date.

***Other subsets of the coal sample were not tested due to limited duplicate units or difficulties in identifying them as such.

****AEP, Penelec, TVA, and Comm Ed together account for 62% of the unit-years for 600+ Mw coal units in the data base. Utility comparisons for smaller units are not feasible because of (i) the much greater dispersion of smaller units among utility operators—86 operating utilities are represented in the coal data base; (ii) more variation in load-following among smaller units which may mask differences in operating reliability.

TABLE 6.1: PROJECTED AVERAGE CAPACITY PERFORMANCE OF HYPOTHETICAL COAL-FIRED UNITS IN FIRST 10 YEARS OF OPERATION†

	Subcritical Units		Supercritical Units	
Unit Size	**FGD**	**No FGD**	**FGD**	**No FGD**
200 Mw	69%	73%	—	—
400 Mw	57-61%	60-65%	59-67%	62-71%
600 Mw	59-62%	62-65%	63-66%	66-70%
800 Mw	54-56%	57-59%	60-63%	63-66%
1000 Mw	49-50%	52-53%	57-63%	60-66%

†Ten-year capacity performance figures are "levelized" rather than arithmetic averages. The 5% scrubber penalty is based on general utility estimates. See Section 8.3. Figures are given in ranges since they are drawn from several capacity performance trend equations. Two equations were used to project figures for 200-Mw units, but only one value is shown since the equations agreed in each case. Figures assume 2% sulfur content and 12.5% ash content and would need adjustment for differences. Broadly, figures increase/decrease by 2% for each percent that coal sulfur content is below/above 2%. Figures are drawn from regression equations based on the entire coal data base, but specifically assume post-1966 "vintage" rather than 1960-66 installation.

These findings are quantified in capacity performance trend equations developed from the CEP data base presented in the remainder of this chapter. These equations assume scrubbers will not be installed on coal-fired units since the data base includes virtually none so equipped. Table 6.1, based on the trend equations, projects capacity performance of hypothetical scrubber-equipped units by deducting 5% from projections for those without.

There are limits to accuracy in projections of future coal capacity factors based on the Table or trend equations.

1. They apply to fully base-loaded plants. Capacity factors would be less for plants which load-follow.

2. Different conventions are sometimes used to define capacity. This leads to capacity factor measurement variations which are sometimes considerable and may mask shifts due to changes in unit size, age, and other variables.*

3. Capacity performance data was computed from capacity factors in the CEP data base by applying average "adjustment factors" for estimated load-following. Factors were developed for size groups of generating units, rather than for each unit in each year.** This tends to mask capacity performance differences related to variables other than unit size. These include age, fuel quality, year of construction, and utility owner's operating patterns.

4. The data base for large (600+ Mw) units is small, 52 units operating for 174 unit-years. Half of this sub-sample falls in the 600-699 Mw range (26 unit, 90 unit-years). Furthermore, over three-fifths of the unit-years for 600+ Mw units are accounted for by only four operating companies (American

*We have overstated capacity compared to definitions employed in other studies (see Appendix B).

**Adjustment factors were developed from the data base of the Equipment Availability Task Force of the Edison Electric Institute. Access to individual unit data in the data base is not permitted by the Task Force, so adjustment factors could not be developed on a less aggregated basis than unit size (see Appendix C).

Electric Power, Commonwealth Edison, Pennsylvania Electric Company, and the Tennessee Valley Authority). Consequently, large coal-fired units in the data base do not provide a complete cross-section.

5. Major changes in regulation, economics, and technology since units in the CEP data base were designed and constructed will affect future capacity performance in ways not now apparent (see Chapter 8).

Section 6.2: Overview of Key Variables Affecting Coal Capacity Performance

Capacity performance of coal-fired units can be correlated to five key variables: *unit size, age, vintage* (year of installation), *boiler "criticality"* (supercritical or subcritical), and *fuel quality*. Statistical trend equations are applied to the entire data base of coal-fired units (100 Mw to 1000+ Mw) later in this section (see Sections 6.3, 6.4., and 6.5 for specific equations for the size groups 100-399 Mw, 400-599 Mw, and 600+ Mw).*

The 100-Mw minimum capacity in the coal data base contrasts sharply to the 450-Mw minimum size in the nuclear data base. Small coal-fired units were included in this study for the following three reasons:

Most U.S. coal units installed during the sample period are in the small (100-399 Mw) range (157 of 250 units). There are only three nuclear units in that range.

Small coal units are still manufactured by power plant vendors, so it is relevant to examine unit records for potential size-based economic superiority.** Nuclear units under 600 Mw are no longer offered by U.S. vendors.

Most experience with coal units after age 10 is derived from those less than 400 Mw, so statistical analysis of plant aging is enhanced by examining small units.

Capacity performance is a better basis than capacity factor for projecting future capacity factors of coal-fired units if correction is made for anticipated load-following. Capacity performance comparisons of individual boiler manufacturers, utility companies, and generating units are not feasible since data relies on average estimates of load-following. It is, therefore, not precise on an individual basis.***

1. *Unit Size:* The most striking finding from the coal unit statistics is that capacity performance is inversely related to unit size. The inverse relationship is strongest below 400 Mw and weakens at higher capacities, as shown in Table 6.2.

*Size groupings were chosen on the basis of similarities observed in trend equations and tabular breakdowns. They parallel size classifications in the EEI Equipment Availability data base.

**The 350-800 Mw range is seen to be the optimal coal unit size in Chapter 9.

***Exceptions are statistical trend equations for capacity performance of large coal-fired units (Section 6.5) in which utility ownership is included as an explanatory variable because of the high concentration of large units among four utilities.

Note that Appendix A provides annual capacity factor data for each coal-fired unit over 400 Mw. Such data for small coal-fired units is available by arrangement with CEP.

TABLE 6.2: COAL-FIRED CAPACITY PERFORMANCE AND CAPACITY FACTOR AS A FUNCTION OF UNIT SIZE

Size Range	Number of Units	Number of Unit-Years	Average Capacity Performance	Standard Deviation	Average Capacity Factor	Standard Deviation
100-199 Mw	67	580	83.3%	13.3%	72.1%	11.5%
200-299 Mw	53	420	76.2	14.5	69.0	13.1
300-399 Mw	37	260	70.4	15.1	65.0	13.9
400-499 Mw	14	76	65.4	16.4	57.1	14.4
500-599 Mw	27	131	63.7	13.1	57.9	11.9
600-699 Mw	26	90	62.1	14.8	57.6	13.7
700-799 Mw	5	27	68.2	12.1	63.2	11.2
800-899 Mw	14	29	62.8	13.3	58.3	12.4
900-999 Mw	5	22	58.3	10.5	54.1	9.8
1000+ Mw	2	6	54.7	10.5	50.7	9.7
Total Coal	250	1641	74.8%	16.1%	66.9%	13.8%

TABLE 6.3: COAL-FIRED CAPACITY PERFORMANCE AND CAPACITY FACTOR BY CEP SIZE GROUPS

Size Group	Number of Units	Number of Unit-Years	Average Capacity Performance	Average Capacity Factor
100-199 Mw	67	580	83.3%	72.1%
200-299 Mw	53	420	76.2%	69.0%
300-399 Mw	37	260	70.4%	65.0%
400-599 Mw	41	207	64.3%	57.6%
600 + Mw	52	174	62.4%	57.9%
TOTAL	250	1641	74.8%	66.9%

Decline in capacity performance and capacity factor with increased size is greatest in the 100-399 Mw range. A plateau in capacity performance is reached mid-range (400-899 Mw). Above 900 Mw, capacity performance again drops off sharply. But the decline is inconclusive as it is based on only seven units operating for 28 unit-years.

Average capacity performance decline per 100 Mw increase in capacity is 1.1 percentage points for supercritical units, 5.6% for subcritical under 400 Mw, 2.2% for subcritical 400+ Mw, and 3.5% for all coal units (see Equations 6.1, 6.3, 6.9, and 6.18). For analysis, it is convenient to group the coal-fired units into five size groups in Table 6.3.

Coal-fired units operate at lower capacity factors in their first full year of commercial operation. They plateau in their second or, for large units, third year. Since large units have operated fewer years in the data base, first and second "immature" years are weighted more heavily for them in Table 6.3. This is shown in Table 6.4.

2. *Unit Age:* Coal-fired units need a full calendar year of commercial operation, two years if large, to reach a stable capacity performance level, as shown in Table 6.5.

Table 6.5 size groups exhibit a "maturation" process to a higher capacity performance in years two through four than in year one. Trend equations indicate average second year capacity performance to be up 2.2% for supercritical units, 7.4% for sub-400 Mw subcritical units, and 5.3% for 400+ Mw subcritical.

Capacity performance begins to decline after several years' operation.* There are two important exceptions, however. Supercritical units generally, and 600+ Mw units of both subcritical and supercritical design, increase capacity performance with increased age. However, rates of increase slow considerably after age two. Moreover, their increased capacity performance is statistically related to their lower average age as compared to other units.

3. *Unit Criticality:* In the late 1950's U.S. utilities installed "supercritical" fossil-fired generating units—stations with boiler operating pressures above

*Apparent increases with increased age in several years for 200-299 Mw and 400-599 Mw units result from higher reliability of earlier installed units and so do not indicate a continuing maturation effect.

TABLE 6.4: COAL-FIRED CAPACITY PERFORMANCE, OMITTING YEARS 1 AND 2 OF OPERATION

	Average Capacity Performance		
Size Range	Including All Years	Omitting Age = 1	Omitting Ages = 1 and 2
100-199 Mw	83.3%	83.9%	83.6%
200-299 Mw	76.2%	77.0%	77.3%
300-399 Mw	70.4%	71.0%	70.5%
400-599 Mw	64.3%	64.6%	64.4%
600 + Mw	62.4%	63.9%	65.4%
TOTAL	74.8%	75.9%	76.3%

3208 lbs/p.s.i. These high pressure units raise average temperature at which steam is fed to the turbine and improve plant efficiency without raising steam temperature above the standard 1050°F which would require use of expensive heat-resistant alloys. Supercritical fossil units reduce fuel costs by higher efficiency, although their average capital cost is slightly higher than that of equal-sized subcritical units.*

The move to supercritical pressure units is coincident with that to large unit sizes. The CEP data base shows supercriticals comprise, by range, none of 100-299 Mw units, 24% of 300-399 Mw, 41% of 400-599 Mw, and 65% of 600+ Mw. The smallest supercritical unit has 346 Mw capacity.

Some utility engineers and planners have suggested that supercritical units are less reliable than subcritical, partially because of their supercritical construction. CEP data base analysis does not support this conclusion.

Capacity performance of supercritical units 300-399 Mw is not statistically different than for subcritical units of comparable size, age, and coal quality. Above 400 Mw, supercritical performance is superior by 4 or more percentage points (see Sections 6.4 and 6.5).

4. *Coal Quality:* Coal burned at U.S. power plants is by no means homogeneous. Coal ash content ranges from 6.0% to 25.9% over 1641 unit-years in the data base. Similarly, sulfur content varies from 0.4% to 6.0%; heat content from 6379 to 13,582 Btu per pound of coal; and moisture content from 2.5% to 38.6%.

Power engineers try to design boilers and accessory equipment to fit specific coal characteristics so as to maximize plant availability to coal grade. Nevertheless, plant capacity performance is positively related to coal ash content at +1.4% per 1% increase in ash content. It is negatively related to sulfur content at −1.8% per 1% increase in sulfur content.** Capacity performance was

*Non-supercritical units are classified by EEI in two groups: "drum-type" and "once-through" subcritical. We have combined them into one "subcritical" category, since 95% of subcritical units 130+ Mw in the EEI data base (1964-73) are drum-type.

**These are averages for the entire coal data base. Subsets of the data base, disaggregated by unit size and boiler criticality, show different coefficients. In some cases they do not have a statistically significant relationship with ash or sulfur content (see Sections 6.3 to 6.5).

Fuel switching, which often results in boilers burning one coal grade when designed for another, could not be tested as an influence on capacity performance.

TABLE 6.5: COAL-FIRED CAPACITY PERFORMANCE AS A FUNCTION OF UNIT AGE, FOR DIFFERENT SIZE GROUPS

Size Group	Age	Number of Unit-Years	Average Capacity Performance
100-199 Mw	1	67	78.8%
	2-4	190	85.5%
	5-7	154	84.2%
	8-10	119	82.9%
	11-13	50	79.3%
	TOTAL	580	83.3%
200-299 Mw	1	53	70.5%
	2-4	152	77.1%
	5-7	112	76.2%
	8-10	71	78.7%
	11-13	32	75.9%
	TOTAL	420	76.2%
300-399 Mw	1	37	67.1%
	2-4	92	73.1%
	5-7	72	71.6%
	8-10	38	66.1%
	11-13	21	68.6%
	TOTAL	260	70.4%
400-599 Mw	1	38	62.9%
	2-4	98	64.7%
	5-7	45	66.3%
	8-10	18	62.6%
	11-13	8	58.6%
	TOTAL	207	64.3%
600+ Mw	1	51	58.9%
	2-4	90	63.1%
	5-7	26	65.3%
	8-10	7	68.4%
	TOTAL†	174	62.4%

†The first 600+ Mw units were installed in 1963, so that no 600+ Mw coal-fired unit had operated for more than 10 full calendar years through 1973, the last year of operation in the CEP coal data base.

TABLE 6.6: COAL-FIRED CAPACITY PERFORMANCE, SUBCRITICAL VS. SUPERCRITICAL†

	Subcritical			Supercritical		
Size Group	Number of Units	Number of Unit-Years	Average Capacity Performance	Number of Units	Number of Unit-Years	Average Capacity Performance
100-199 Mw	67	580	83.3%	0	0	—
200-299 Mw	53	420	76.2%	0	0	—
300-399 Mw	28	194	71.2%	9	66	68.4%
400-599 Mw	24	119	63.8%	17	88	65.0%
600+ Mw	18	66	60.8%	34	108	63.4%
TOTAL	190	1379	76.7%	60	262	65.3%

†Since capacity factor data was converted to capacity performance figures using average load-following data which was not disaggregated by boiler type, Table 6.6 may not be a precise indicator of capacity performance difference between subcritical and supercritical units. But utility engineers report that load-following is not significantly related to boiler type. Subcritical units are more amenable to economy shutdowns, due to easier shutdown and startup. Supercritical units better accommodate partial load swings.

generally not related to coal heat content.*

5. *Vintage:* Units in the CEP data base were installed between January 1, 1960, and December 31, 1972. It is convenient to divide units into three "vintages," or periods of installation—1960-63, 1964-66, and 1967-72. The earliest vintage accounts for 59% of all subcritical unit-years. In contrast supercritical units tend to have larger Mw capacity. Their more frequent installation in recent years helps place 67% of their unit-years in the recent 1967-72 vintage. Only 18% of supercritical unit-years are 1960-63 vintage.

Trend equations suggest subcritical coal units of 1967-72 vintage operate at a capacity performance level 5.4% less than comparable units of 1964-66 vintage, and 11.3% less than those of 1960-63. Supercritical unit vintage was not correlated with capacity performance in a statistically significant way.

6. *The Capacity Performance Trend Equations:* Following are correlation equations, aggregated for all unit sizes.

*(Equation 6.1**)*
All Coal Unit-Years (N = 1627)

$$\text{Capacity Performance} = \underset{}{81.05} - \underset{(0.232)}{3.479} \times (\text{Mw}/100) + \underset{(1.00)}{11.60} \times \text{V60}$$

$$+ \underset{(1.05)}{4.38} \times \text{V64} - \underset{(1.07)}{6.56} \times \text{XA} - \underset{(0.13)}{1.05} \times \text{AGE}$$

$$+ \underset{(0.34)}{3.55} \times \text{XASH} - \underset{(0.13)}{1.80} \times \text{XSUL} + \underset{(1.14)}{2.65} \times \text{CR}$$

(Adjusted $R^2 = 0.33$)

*Coal moisture content could not be tested in the trend equations because of its high "collinearity" with heat content (moisture content varies oppositely with heat content).

**The variables on the right-hand side of the equation are as follows:

Mw = unit capacity in megawatts

V60 is a "dummy variable" for unit vintage, whose value is 1 if the unit went commercial during 1960-63, and zero otherwise.

V64, similarly, equals 1 if the unit went commercial during 1964-66, and zero otherwise.

XA is a dummy variable representing the first full year of commercial operation. It is equal to 1 when AGE = 1, and zero otherwise.

AGE = Unit age

XASH is a variable representing the average ash content of the coal burned by each unit in a particular year. Its value is as follows:

XASH = 1 if coal ash content is less than 10%
XASH = 2 if coal ash content is between 10% and 12.5%
XASH = 3 if coal ash content is between 12.5% and 15%
XASH = 4 if coal ash content is greater than 15%

(Ash content is respresented in this fashion in the regression equations, rather than being represented directly, in order to reduce the impact of very small or large ash contents. Most ash contents are in the 10%-15% range.)

XSUL is a variable representing the coal sulfur content. Its value is as follows:

XSUL = 1 if coal sulfur content is less than 1%
XSUL = 2 if coal sulfur content is between 1% and 2%

(Equation 6.2)
All Subcritical Coal Unit-Years (N = 1366)

$$\text{Capacity Performance} = 82.67 - \underset{(0.255)}{4.258} \times (\text{Mw}/100) + \underset{(1.06)}{11.33} \times \text{V60}$$

$$+ \underset{(1.13)}{5.36} \times \text{V64} - \underset{(1.20)}{8.02} \times \text{XA} - \underset{(0.14)}{1.11} \times \text{AGE}$$

$$+ \underset{(0.36)}{3.84} \times \text{XASH} - \underset{(0.30)}{1.70} \times \text{XSUL}$$

(Adjusted $R^2 = 0.33$)

(Equation 6.3)
All Supercritical Coal Unit-Years (N = 261)

$$\text{Capacity Performance} = 75.15 - \underset{(0.481)}{1.067} \times (\text{Mw}/100) - \underset{(1.10)}{1.82} \times \text{AGED}^*$$

(Adjusted $R^2 = 0.04$)

Equations 6.1 and 6.2 are heavily weighted by small subcritical units. They account for 61% of Equation 6.1 unit-years and 84% in Equation 6.2. Therefore, the equations are not useful for projecting capacity performance of supercritical or large subcritical units, although they are useful in discerning trends related to coal quality and vintage. Equation 6.3 yields values for supercritical units consistent with others derived from CEP data base trend equations (see Equations 6.12 and 6.13).

Following is the levelized average supercritical capacity performance for the first 10 years of operation (based on Equation 6.3).

(Equation 6.4)

$$\text{Capacity Performance} = 72.08 - 1.067 \times (\text{Mw}/100)$$

XSUL = 3 if coal sulfur content is between 2% and 3%
XSUL = 4 if coal sulfur content is between 3% and 4%
XSUL = 5 if coal sulfur content is greater than 4%

CR is a variable representing boiler criticality. Its value is zero for subcritical units, one for supercritical units.

The figures in parentheses under the coefficients are "standard error coefficients" (see Section 3.3 and Appendix F).

*AGED is a function of unit age—AGED = 1/Log (1 + AGE). A negative AGED term indicates increased capacity performance as age increases, with the greatest rate of increase for small AGE values (see Section 3.3).

Other variables were not statistically significant in the correlation equation for all supercritical units.

This equation is employed in Sections 6.3, 6.4, and 6.5, below.*

Section 6.3: Performance of Coal-Fired Units 100-399 Mw

The 157 units in this size group operated for a total of 1260 unit-years in 1961-1973, or 77% of the unit-years in the CEP coal data base. This size group is heavily weighted with very small, older units, the superior capacity performance of which accounts for the high 78.3% average of the whole (see Table 6.7).

A strong relationship between capacity performance, unit size, and vintage is apparent in the regression equation for capacity performance of 100-399 Mw units.

(Equation 6.7)

$$\text{Capacity Performance} = \underset{}{84.95} - \underset{(0.460)}{5.497} \times (\text{Mw}/100) + \underset{(1.15)}{12.22} \times \text{V60}$$

$$+ \underset{(1.26)}{5.70} \times \text{V64} - \underset{(1.29)}{7.98} \times \text{XA} - \underset{(0.14)}{1.03} \times \text{AGE}$$

$$+ \underset{(0.37)}{3.36} \times \text{XASH} - \underset{(0.32)}{1.55} \times \text{XSUL}$$

Adjusted $R^2 = 0.26$
N = 1248

Equation 6.7's implications for capacity performance of small coal-fired units are as follows:

1. *Unit Size:* Capacity in Mw is negatively related to capacity performance at 5.5 percentage points per 100 Mw increment from 100-399 Mw.
2. *Vintage:* Units installed during 1964-66 average 5.7 percentage points higher than units installed 1967-72, apart from the effect of other variables. Similarly, units of 1960-63 vintage average 12.2 points higher than comparable 1967-72 units, and 6.5 points higher (12.2 minus 5.7) than 1964-66 units.

*10-year levelized average versions of Equations 6.1 and 6.2 are:

(Equation 6.5)

All units: Capacity Performance = 79.49 − 3.479 × (Mw/100) + 2.65 × CR

(Equation 6.6)

All subcritical units: Capacity Performance = 81.56 − 4.258 × (Mw/100)

These equations assume 1967-72 vintage (V60 = V64 = 0), 2% sulfur content (XSUL = 2.5), 12.5% ash content (XASH = 2.5).

TABLE 6.7: CAPACITY PERFORMANCE OF SMALL COAL-FIRED UNITS BY UNIT SIZE AND YEAR OF INSTALLATION

	100-199 Mw		**200-299 Mw**		**300-399 Mw**		**100-399 Mw**	
1960-73	85.2%	*406*	80.1%	*240*	71.2%	*132*	81.3%	*778*
1964-66	80.6%	*117*	71.3%	*98*	73.1%	*70*	75.6%	*285*
1967-72	75.2%	*57*	70.7%	*82*	65.4%	*58*	70.4%	*197*
TOTAL	83.3%	*580*	76.2%	*420*	70.4%	*260*	78.3%	*1260*

The numbers in italics are the number of unit-years.

3. *Age:* 100-399 Mw units reach highest capacity performance in the second full year of operation, at seven percentage points more than first-year capacity performance. They decline 1% per year through age 13, the oldest year of operation in the coal data base.*

4. *Coal Quality:* Capacity performance drops relative to sulfur content at 1.6 percentage points for each percent of sulfur content. It climbs, relative to ash content, at 1.3% for each percent of ash content.**

5. *Unit Criticality:* Since subcritical units account for 95% of 100-399 Mw unit-years, boiler criticality is not significant in the statistical analysis.

These observations are discussed in Chapter 7.

Equation 6.7 may be made more tractable for projecting hypothetical capacity performance by assuming average coal quality (2% sulfur content, so XSUL = 2.5, and 12.5% ash content, so XASH = 2.5) and post-1966 vintage, and taking a 10-year levelized average, in Equation 6.8.

*(Equation 6.8***)*

$$\text{Capacity Performance} = 83.43 - 5.497 \times (\text{Mw}/100)$$

Equation 6.8 may then be used to project capacity performance of new coal-fired units (100-399 Mw) (see Table 6.8).

Table 6.8 assumes capacity performance of future small coal-fired units will likely follow those of recent vintage (1966-72). An additional 5.7 percentage points would be added for 1964-66 vintage and 12.2% for 1960-63 vintage (see Equation 6.7). However, as will be shown, it is unlikely that capacity performance trends for small units of earlier vintage will be repeated for new units. Similarly, changes in assumed ash and sulfur content would be reflected in Table 6.8 by application of coal quality coefficients in Equation 6.7. Finally,

*The figure of seven percentage points was computed as: 7.98 times 1, minus 1.03 times 2, plus 1.03 times 1, which equals 6.95.

**1.3% was computed by dividing the 3.36 coefficient of XASH by 2.5, which is the percent change in ash content per unit change in XASH for most ash content variables (see Equation 6.1).

***The 10-year levelized average value of $-7.98 \times XA - 1.03 \times AGE$ in Equation 6.7 is computed to be -6.05. The terms, $3.36 \times XASH - 1.55 \times XSUL$ become 4.53 for the coal quality assumed above (2% sulfur, 12.5% ash). Then the intercept 84.95 is reduced by $6.05 - 4.53$ to become 83.43.

TABLE 6.8: PROJECTED 10-YEAR AVERAGE CAPACITY PERFORMANCE OF SMALL COAL-FIRED UNITS FROM EQUATION 6.8

Unit Capacity	Projected Capacity Performance†
150 Mw	75.2%
250 Mw	69.7%
350 Mw	64.2%

†Equation 6.4, which gives a 10-year levelized average capacity performance, based on all supercritical units (Capacity Performance = 72.08 − 1.067 × (Mw/100), yields a capacity performance value of 68.3% for 350-Mw units. Equation 6.4 is a better indicator of small supercritical capacity performance than Equations 6.7 or 6.8, since 25% of its unit-years are provided by such units, against 5% in Equations 6.7 and 6.8.

Equation 6.4 should be a reliable indicator for capacity performance of small subcritical units, since such units contribute 95% of its sample space. Alternatively, Equation 6.9, based only on subcritical units 100-399 Mw, yields capacity performance values about 0.5% higher than those in Table 6.8.

(Equation 6.9)

$$\text{Capacity Performance} = \underset{}{85.18} - \underset{(0.492)}{5.580} \times (\text{Mw}/100) + \underset{(1.18)}{11.60} \times \text{V60}$$

$$+ \underset{(1.30)}{5.90} \times \text{V64} - \underset{(1.32)}{8.48} \times \text{XA} - \underset{(0.14)}{1.08} \times \text{AGE}$$

$$+ \underset{(0.38)}{3.79} \times \text{XASH} - \underset{(0.32)}{1.68} \times \text{XSUL}$$

(Adjusted $R^2 = 0.25$, N = 1182)

use of scrubbers would reduce figures in Table 6.8 by about 5% (three to four percentage points).

Despite Table 6.8's high record for 100-300 Mw coal-fired units, their small capacity leaves the economic edge with large units because of their much lower per-unit capital costs (see Chapter 9).

Section 6.4: Performance of Coal-Fired Units 400-599 Mw

The 41 units in this size group operated for 207 unit-years in 1961-73 at an average capacity performance of 64.3%. Variations in this size group are best explained by Equation 6.10.

(Equation 6.10)

$$\text{Capacity Performance} = \underset{}{65.73} + \underset{(2.88)}{11.97} \times V60 + \underset{(2.50)}{8.56} \times V64$$

$$- \underset{(0.41)}{1.30} \times AGE + \underset{(1.13)}{2.82} \times XASH$$

$$- \underset{(1.15)}{3.23} \times XSUL + \underset{(2.01)}{3.76} \times CR$$

Adjusted $R^2 = 0.14$
$N = 206$

Equation 6.10 shows strong similarities between the 100-399 Mw class and that of 400-599 Mw in capacity performance to date.

1. *Vintage:* Post-1966 units average 8.6% less than comparable units installed in 1964-66 and 12.0% less than the 1960-63 vintage. This is similar to vintage effects for 100-399 Mw units in Equation 6.7.

2. *Age:* 400-599 Mw units averaged 1.3 percentage point drop for each advancing year of operation, similar to the age effect for 100-399 Mw units. Unlike them, however, 400-599 Mw units did not have a statistically significant lower capacity performance during their first full calendar year of commercial operation. Medium-size coal units have not required a maturation period to reach their highest capacity factors.

3. *Coal Quality:* Capacity performance drops at about 3.2% per 1% of sulfur content and climbs about 1.1% per 1% of ash content. The ash effect is comparable to small units, while sulfur content effect is twice that observed.*

Unit size is not statistically significant in capacity performance variations within the 400-599 Mw size group. But it does influence overall capacity level of 400-599 Mw units in that trend equations yield lower capacity performance values for 400-599 Mw units than for smaller units. This is shown by Equation 6.11 which gives the 10-year levelized average capacity performance of 400-599 Mw coal-fired units burning an average grade of coal (2% sulfur, 12.5% ash).

(Equation 6.11)

$$\text{Capacity Performance} = 58.56 + 11.97 \times V60 + 8.56 \times V64 + 3.76 \times CR$$

For medium-size coal units of current (post-1966) vintage, burning average

*Capacity performance for 400-599 Mw units increases about 3.4 percentage points per 1000 Btu/lb increment in heat content of the coal burned. However, the equation used here for 400-599 Mw units (Equation 6.10) was developed without using coal heat content as a variable in order to maintain consistency with 100-399 Mw units. Heat content was not a statistically significant variable for any regression equations for any size group other than 400-599 Mw.

TABLE 6.9: UTILITIES OPERATING LARGE COAL-FIRED UNITS

	Number of Units/Number of Unit Years		
Utility	**Subcritical**	**Supercritical**	**All Units**
AEP	0/0	8/28	8/28
Comm Ed	5/27	0/0	5/27
Penelec	0/0	6/24	6/24
TVA	2/20	2/9	4/29
Other Utilities	11/19	18/47	29/66
Total	18/66	34/108	52/174

coal, Equation 6.11 indicates that the average 10-year levelized capacity performance would be 58.6% for subcritical units and 62.3% for supercritical units.* These figures do not reflect the potential impact of scrubbers on capacity performance (probably 5% or three percentage points) (see Chapter 8).

Section 6.5: Performance of Coal-Fired Units 600 Mw or Larger

The 52 units in this size group have operated for a total of 174 unit-years at an average capacity performance of 62.4%: 63.4% for supercritical units (108 unit-years), 60.8% for subcritical units (66 unit-years).

Large unit (600+ Mw) capacity performance is analyzed in more detail than small because (i) debate over coal and nuclear capacity factors focuses on large units of each type, in part because many coal-fired units and all nuclear units planned by U.S. utilities are 600+ Mw due to economies of scale realized in large unit capital cost; (ii) the CEP data base is more homogeneous for large than small coal units—14 out of 18 large subcritical units and all of the 34 large supercritical units were installed in 1967 or later, eliminating the vintage effect encountered with smaller units.**

Major statistical conclusions drawn about large coal units are:

a) Higher supercritical capacity performance when compared to subcritical units.

b) Unit size negatively related to capacity performance, with the decline more pronounced for subcritical units when compared to supercritical.

c) Increased capacity performance with advancing unit age. This conclusion is limited by the absence of substantial large subcritical unit experience and any large unit supercritical experience beyond age six.

d) Coal quality not a significant variable, except that sulfur content is negatively related to capacity performance for supercritical units. This finding is not attributable to scrubbers, since no large coal unit in the CEP data base has operated with scrubbers during the sample period.

*The 58.6% figure for subcritical units is consistent with Equation 6.6 for all subcritical units, but is considerably less than indicated by Equation 6.17, discussed below. Similarly, the 62.3% figure for supercritical units is substantially less than indicated by Equation 6.4 which is based on all supercritical units.

**The first full calendar years of operation for the four earlier large subcritical units are 1964 (two units), 1966, and 1967.

TABLE 6.10: CAPACITY PERFORMANCE OF LARGE SUPERCRITICAL COAL UNITS AS A FUNCTION OF UNIT AGE

Unit Age	Number of Unit Years	Average Capacity Performance
1	37	60.3%
2	30	62.2%
3	18	68.3%
4	13	64.2%
5	6	67.0%
6	4	71.8%
Total	108	63.4%

The data indicates the percentage of levelized average 10-year capacity performance for subcritical units is in the mid-60's for 600-Mw units and significantly less for larger. The percentage for supercritical units would be in the high-60's for 600-Mw units and slightly less for larger.

Large unit analysis is rendered more complex by operating experience concentration among only four U.S. utilities. Together, American Electric Power Company (AEP), Commonwealth Edison of Chicago (Comm Ed), Pennsylvania Electric Company (Penelec), and Tennessee Valley Authority (TVA) account for 108 (62%) of 174 unit-years for large coal units in the CEP data base as shown in Table 6.9.

The four utilities account for 71% of subcritical and 56% of supercritical unit-years for large coal units. By comparison, the four nuclear utilities with the most unit-years (Commonwelath Edison, Northeast Utilities, Southern California Edison, and Wisconsin Electric Power) together account for only 41% of BWR/PWR unit-years in the CEP data base (see Table 3.5). Company concentration of coal units under 600 Mw is considerably less than that of either large coal or nuclear. Concentration of large coal unit experience among so few companies complicates analysis since company operation must be considered as a variable.

6.5a: Large Supercritical Units

The two key trend equations for large supercritical units are as follows:

(Equation 6.12)

$$\text{Capacity Performance} = 94.95 - \underset{(0.922)}{2.394} \times (\text{Mw}/100) - \underset{(1.59)}{3.45} \times \text{AGED}^{*} - \underset{(1.07)}{1.94} \times \text{XSUL}$$

Adjusted $R^2 = 0.11$
N = 108

*A negative "AGED" term indicates higher capacity performance with increasing age (see Equation 6.3, Section 6.2).

(Equation 6.13)

$$\text{Capacity Performance} = 77.95 - \underset{(1.45)}{3.60} \times \text{AGED}^{*} - \underset{(2.98)}{10.50} \times \text{PENN}^{**} + \underset{(2.90)}{9.27} \times \text{AEP}^{**} - \underset{(1.01)}{2.34} \times \text{XSUL}$$

Adjusted $R^2 = 0.29$
N = 108

Equation 6.12 is that of best-fit for large supercritical units when "company" variables are excluded *(i.e.,* when only variables for unit size, age, vintage, and coal quality are tested). Equation 6.13 is that of best-fit when all variables, including company operation, are tested.

Equations 6.12 and 6.13 yield the following conclusions for large supercritical coal units:

1. *Unit Age:* Capacity performance correlates with unit age by 10th year projection increases of about eight percentage points over the first year, based on terms "−3.45 × AGED" and "−3.60 × AGED" in Equations 6.12 and 6.13. The extent of maturation is slightly less than the average projected 10 percentage point increase from age one to age 10 for nuclear PWR's indicated by the "−4.22 × AGED" term in Equation 3.1. This relationship is shown in Table 6.10.

Average capacity performance of units age three and older is 67.1%, six percentage points higher than the 61.2% average for ages one to two.**

2. *Coal Quality:* Equations 6.12 and 6.13 both indicate an average 2% capacity performance decline for each 1% increment in coal sulfur content. This decline is analogous to the inverse relationship between sulfur content and capacity performance for 100-399 Mw and 400-599 Mw units. However, ash content was not statistically related to capacity performance for large supercritical units.

3. *Company Operation:* Equation 6.13 suggests the 24 Penelec unit-years were at capacity performance levels 10½ percentage points less than the 84 others, correcting for age, size, and sulfur content. Conversely, AEP operation ran 9¼% higher than comparable units. This conclusion is reflected in Table 6.11.

*PENN is a "company" variable which is set equal to one for a unit operated by Penelec and zero otherwise. The company variables AEP (which also appears in Equation 6.13), COMM and TVA (which appear in regression equations for subcritical units) are defined analogously. TVA operates nine large supercritical unit-years in the CEP data base but was not a statistically significant variable for such units.

**While Table 3.13 may overstate the maturation of large supercritical units by masking unit size differences (since units at the smaller end of the 600+ Mw range are more heavily represented among older unit-years), Equations 6.12 and 6.13 indicate a clear maturation tendency.

TABLE 6.11: CAPACITY FACTORS OF LARGE SUPERCRITICAL COAL UNITS BY UTILITY OPERATOR

Utility	Number of Units/ Unit-Years	Average Capacity	Average Capacity Factor
AEP	8/28	694 Mw	66.7%
Penelec	6/24	844 Mw	49.1%
Other	20/56	772 Mw	59.1%
All Units	34/108	768 Mw	58.8%

While use of actual capacity factor data alone does not correct for differences in unit size, unit age, or sulfur content, Table 6.11 makes it clear that AEP units operate at notably higher capacity factors than their counterparts and Penelec units' record is poor* (see Chapter 7).

4. *Unit Size:* Mw capacity in Equation 6.12 decreases at 2.4% capacity performance per 100-Mw increment. However, the statistically significant relationship between unit size and capacity performance is eliminated when company variables are introduced in Equation 6.13.** This is because the low capacity performance of Penelec unit-years, which average 844 Mw capacity, and high record of AEP (694 Mw), are substantially explained in Equation 6.13 by company variables rather than by the unit size variable in Equation 6.12.

And so the relationship between unit size and capacity performance is ambiguous for large supercritical units. It is significantly negative in one case and, though still negative, not statistically significant in the other. However, the trend equation for all supercritical units (Equation 6.4) also has a statistically significant inverse relationship between unit size and capacity performance.

Equations 6.12 and 6.13 may be converted into levelized 10-year average capacity performance equations, assuming an average sulfur content of 2% (XSUL = 2.5), as done in Equations 6.14 and 6.15.

(Equation 6.14)
(from Equation 6.12)

$$\text{Capacity Performance} = 84.27 - 2.394 \times (\text{Mw}/100)$$

(Equation 6.15)
(from Equation 6.13)

$$\text{Capacity Performance} = 66.02 - 10.50 \times \text{PENN} + 9.27 \times \text{AEP}$$

*Note that the capacity factor data in Table 6.11 slightly understates capacity performance, which is the index of operating reliability used for coal. Further, neither capacity factor nor capacity performance data incorporate potential effect of different amounts of load-following by different utilities' units. But it is unlikely that the differences in capacity factor can be explained by differences in load-following.

**The Mw/100 term in the regression equation, when company variables are included, is negative 1.19. However, the stadard error term is 1.16. So the negative 1.19 coefficient is significant only to the 85% confidence level, less than the 95% level used as a threshold for significance in this study.

Equations 6.14 and 6.15 are comparable to Equation 6.4, which gives 10-year average capacity performance values, from the sample of all supercritical units.

(Equation 6.4)
(from Equation 6.3)

$$\text{Capacity Performance} = 72.08 - 1.067 \times (\text{Mw}/100)$$

Table 6.12 shows the three relevant trend equations for large supercritical units to yield levelized capacity performance values in the same approximate range for each unit size: 66%-70%/600 Mw, 66%/800 Mw, 60%-66%/1000 Mw. These figures do not incorporate scrubber effect, estimated to reduce capacity performance by 5% (three to four percentage points).

6.5b: Large Subcritical Units

Equation 6.16 shows the best fit for large subcritical units.

(Equation 6.16)

$$\text{Capacity Performance} = 91.23 - \underset{(1.58)}{3.284} \times (\text{Mw}/100) - \underset{(1.73)}{3.60} \times \text{AGED}$$

$$- \underset{(3.32)}{9.86} \times \text{COMM} + \underset{(3.67)}{9.55} \times \text{TVA}$$

Adjusted $R^2 = 0.41$
N = 66

Coal quality variables (sulfur, ash and heat content) were not statistically significant when tested with company operation variables in Equation 6.16.* The equation yields the following about performance of large subcritical units:

1. *Unit Age:* Tenth-year capacity performance would be slightly over eight percentage points higher than that of the first year, based on the term "$-3.60 \times$ AGED" in Equation 6.16. This maturation is identical to that for large supercritical units in Equation 6.13. The trend toward higher capacity performance with increasing age for large subcritical units is shown in Table 6.13.

Average capacity performance for units age 3+ is 63.7%—7.7% higher than the 56.0% average for ages one to two.

2. *Unit Size:* Capacity performance drops by 3.3% per 100 Mw increment. This is equal to the unit size "penalty" for nuclear units in Equations 3.3 and 3.1.

*Equation 6.16, absent the company ownership variables for Commonwealth Edison and TVA, is remarkably similar to Equation 3.1 for PWR capacity factors: Capacity Factor = 93.48 − 3.368 × (Mw/100) − 4.22 × AGED.

Ten-year average levelized capacity performance for large subcritical coal units from Equation 6.16 is almost identical to average levelized capacity factor for nuclear PWR's in Equation 3.2.

TABLE 6.12: 10-YEAR LEVELIZED AVERAGE CAPACITY PERFORMANCE FOR LARGE SUPERCRITICAL UNITS, FROM DIFFERENT TREND EQUATIONS (Excluding effect of Company Variables)

Unit Size	Equation 6.14	Equation 6.15	Equation 6.4
600 Mw	69.9%	66.0%	65.7%
800 Mw	65.1%	66.0%	63.5%
1000 Mw	60.3%	66.0%	61.4%
Adjusted R^2-value	0.11	0.29	0.04

TABLE 6.13: CAPACITY PERFORMANCE OF LARGE SUBCRITICAL COAL UNITS AS A FUNCTION OF UNIT AGE

Unit	Number of Unit-Years	Average Capacity Performance
1	14	55.3%
2	11	56.9%
3	10	62.3%
4	8	62.6%
5	7	63.8%
6	5	59.4%
7-8	7	66.7%
9-10	4	69.0%
Total	66	60.8%

TABLE 6.14: CAPACITY FACTORS OF LARGE SUBCRITICAL COAL UNITS BY UTILITY OPERATOR

Unitity	Number of Units/ Unit-Years	Average Capacity	Average Capacity Factor†
Comm Ed	5/27	669 Mw	49.0%
TVA	2/20	704 Mw	67.0%
Other	11/19	695 Mw	55.7%
All Units	18/66	687 Mw	56.4%

†Note that the capacity factor data in Table 6.14 slightly understates capacity performance. As noted in Table 6.11, it is unlikely that the capacity factor differences in Table 6.14 are attributable to different amounts of load-following by different utilities.

TABLE 6.15: 10-YEAR LEVELIZED AVERAGE CAPACITY PERFORMANCE FOR LARGE SUBCRITICAL UNITS, FROM DIFFERENT TREND EQUATIONS (Excluding Effect of Company Variables)

Unit Size	Equation 6.17	Equation 6.18
600 Mw	65.4%	61.9%
800 Mw	58.9%	57.6%
1000 Mw	52.3%	53.2%
Adjusted R^2-value	0.41	0.16

3. *Company Operation:* Equation 6.16 shows units operated by Commonwealth Edison to have 10% less capacity performance than comparable units of other utilities. TVA units operated at 9½% more than other units. This conclusion is reflected in Table 6.14.

A levelized average 10-year capacity performance for large subcritical units, based on the CEP data base, can be obtained from Equation 6.17 (derived from Equation 6.16).

(Equation 6.17)

$$\text{Capacity Performance} = 85.15 - 3.284 \times (\text{Mw}/100) - 9.86 \times \text{COMM} + 9.55 \times \text{TVA}$$

If the unit is assumed to be operated by a utility other than Commonwealth or TVA, the last two terms in Equation 6.17 drop out, giving:

$$\text{Capacity Performance} = 85.15 - 3.284 \times (\text{Mw}/100)^{*}$$

This equation may be compared to Equation 6.18 which is drawn from analysis of all subcritical units 400 Mw or larger.

*(Equation 6.18**)*

$$\text{Capacity Performance} = 74.84 - 2.160 \times (\text{Mw}/100)$$

*This relationship is close to the predicted levelized average capacity factor for nuclear PWR's (Equation 3.2): Capacity Factor = 86.38 − 3.368 × (Mw/100), yielding capacity performance figures for large subcritical coal units about 0.5% less than the capacity factor figures for PWR's of the same unit size.

**Equation 6.18 is a 10-year levelized average version of Equation 6.19 for subcritical units 400+ Mw:

$$\text{Capacity Performance} = 82.64 - \underset{(0.914)}{2.160} \times (\text{Mw}/100) + \underset{(2.82)}{10.57} \times \text{V60} - \underset{(2:92)}{6.62} \times \text{XA} - \underset{(0.49)}{1.37} \times \text{AGE} + \underset{(1.10)}{4.38} \times \text{XASH} - \underset{(1.08)}{4.52} \times \text{XSUL}$$

Adjusted $R^2 = 0.16$
N = 184

TABLE 6.16: CAPACITY PERFORMANCE FOR LARGE COAL-FIRED UNITS WITHOUT SCRUBBERS, FROM TREND EQUATIONS

Unit Size	Supercritical Units	Subcritical Units
600 Mw	66-70%	62-65%
800 Mw	63-66%	57-59%
1000 Mw	60-66%	52-53%

From Table 6.15, the two trend equations for large subcritical units yield levelized capacity performance values which are in fairly close agreement for each unit size: 62%-65%/600 Mw, 57%-59%/800 Mw, 52%-53%/1000 Mw (without scrubbers).

6.5c: Large Subcritical and Supercritical Units Compared

Trend equations for large (600+ Mw) coal-fired units yield the average 10-year levelized capacity performance figures in Table 6.16 for hypothetical coal-fired units of different capacities, without scrubbers. Use of scrubbers would reduce Table 6.16 figures by 5%, or about three percentage points (see Chapter 8).

Table 6.16 figures do not reflect operation by AEP, Commonwealth Edison, Penelec, or TVA. They also assume a 2% sulfur content. While sulfur content is not significantly correlated to capacity performance for large subcritical units, it is negatively correlated for large supercritical units. Based on Equations 6.12 and 6.13, values in the left column of Table 6.16 should be adjusted by 2%-2.3% for each 1% variation in coal sulfur content from the assumed level.

Large supercritical units can be expected to operate at a higher capacity performance than large subcritical units. Differences of several percentage points for 600-Mw units will increase, however, to 6%-8% at 800-Mw and 10% at 1000 Mw. The difference will be more for coal with less than 2% sulfur content and less for 2%+. This suggests the more reliable design choice for large new coal-fired units to be supercritical. In fact, 34 (65%) of the 52 large coal units in the CEP data base operate with supercritical boilers and 34 of the 48 large coal units installed after 1966 are supercritical.

The average capacity of large coal unit-years in the data base is 768 Mw for supercritical and 687 Mw for subcritical. Trend equations project ten-year average levelized capacity performance for units of such capacity of 66% (supercritical) and 63% (subcritical) if impact of 400-599 Mw units is excluded. These figures are slightly higher than actual large unit average capacity performance to date, which is 63.4% (supercritical) and 60.8% (subcritical). The higher figures result from extrapolating, through age 10 (with Equations 6.13, and 6.16), the trend towards increased capacity performance with advancing age shown by large coal units.

A similar assumption was made in applying Equation 3.1, which includes a term for "maturation," to project future capacity factors for PWR's. The extrapolation of a maturing tendency for coal-fired units is only slightly more grounded in actual experience than the similar extrapolation for nuclear units. The average age of large coal units in the CEP data base (3.3 years per unit) is only marginally greater than that for nuclear units (2.8 years per unit).

7: Factors Affecting Coal-Fired Plant Capacity Performance

Section 7.1: Key Findings

1. Lower capacity performance of recent sub-600 Mw subcritical coal units is partially due to reduced capability margins causing failure to meet design capacity requirements.

2. Sulfur content reduction of capacity performance results from corrosion by sulfur compounds formed in coal combustion. Ash effect on capacity performance is inversely related to coal ash-softening temperature, but can be minimized by proper match of boiler design to coal grade.

3. Lower capacity performance of large coal units results from more components, more demanding service conditions, or larger components more difficult to assemble and repair. The smaller effect of unit size on coal than on nuclear performance is partially due to easier regulation permitting operation of coal plants despite equipment breakdown.

4. Supercritical boilers have operated at higher levels than subcritical, despite more complex design.

5. Average decline in coal capacity performance in Ages 10-13 is only 1% per year, compared to decline several times greater for three prototype nuclear plants after Age 10.

6. Superior performance of AEP large units is a product of conservative design practices and extrapolation from smaller units. Pennsylvania Electric's poor large unit capacity performance results from inexperience with large units and inadequate boiler design margins.*

*Full explanation of these findings would require a detailed engineering analysis of the design, construction, and operation of modern coal-fired units. Such an analysis is not now available in the technical literature on power engineering. EPRI classifies causes of nuclear plant capacity reductions (see Chapter 4), but has not published a comparable classification for fossil units.

The major work in this area is the set of three engineering consultant reports on power plant reliability prepared for the Federal Energy Administration (see Bibliography). Each of these reports, however, deals only with two coal-fired and two nuclear plants, and does not address broad conclusions identified here. Similarly, many papers on plant availability presented in industry journals and conferences report on operating problems at specific units or classes of units, but decline to describe or explain overall reliability trends.

Section 7.2: Capacity Measurements and Vintage Effects

Trend equations indicate a significant vintage effect for subcritical coal 100-599 Mw units. Units constructed during 1967-72 have a capacity performance six to eight percentage points less than 1964-66 units, and 12% less than those of 1960-63. Accordingly, figures projected for hypothetical coal-fired units in the tables do not directly reflect the superior capacity performance of units of pre-1967 vintage (see Chapter 6).*

Utility engineers indicate that a form of up-rating has apparently occurred for newer fossil units. Some units have been assigned nameplate ratings which are equal to or greater than their actual generating capability. Formerly, a nameplate rating was assigned that equaled or was less than unit capability. Power plant engineers and manufacturers are now less likely to overbuild, as they did in the past, to ensure the actual generating capability of the unit meets the design specification of the utility, the basis of nameplate rating. "Nameplate rating today doesn't mean the same as it used to," advises one utility engineer.[1]

To the extent that this definition difference has taken place, new units operate at lower capacity factors than their predecessors since they are rated higher and so need to generate more electricity to achieve the same capacity factor. It remains to be seen whether lower capacity performance of new, small subcritical units is due only to difference in capacity definition, or whether design, construction, and operating practices have deteriorated.

Smaller design margin for new units may be cost-effective in some instances.** But this tendency does suggest that capacity performance projections for new coal-fired units should be weighted toward recent data since it is likely to more closely resemble capacity definitions of the recent past. The CEP capacity performance projections assume 1967-72 vintage where vintage was a significant variable in the trend equations.

Section 7.3: Coal Quality

Relationships between capacity performance and coal quality variables observed in the trend equations (see Chapter 6) are summarized in Table 7.1. It shows coal sulfur content negatively related to capacity performance and ash content positively related for units under 600 Mw. Heat content was positively

*It is assumed that new subcritical 100-599 Mw units tend to operate at a vintage effect commensurate with the capacity performance levels of 1967 and later units. Therefore, "V60" and "V64" are equal to zero in developing capacity performance estimates from the trend equations for this sample of units. However, the relationships between capacity performance and explanatory variables of unit size, age, fuel quality, and boiler type in Chapter 6 are based on all data for sub-600 Mw units and not just data for units installed in 1967 or later.

Above 600 Mw only 4 of 18 subcritical units and none of 34 supercritical units were installed prior to 1967. Since the capacity performance of these earlier units was not statistically different from that of later units, vintage effect was not observed for large units (or for supercritical units of all unit sizes).

**Economic criterion of power plant planning is total cost per kwh. Higher capacity factors are not cost-effective if they are purchased by increases in capital investment which raise total cost of kwh's generated.

TABLE 7.1: EFFECT OF COAL QUALITY ON CAPACITY PERFORMANCE

Size Group	Percentage Change in Capacity Performance per Unit Change in Coal Quality Variable†	
	Sulfur Content	**Ash Content**
100-399 Mw	-1.6%	+1.3%
400-599 Mw	-3.2%	+1.1%
600+ Mw Supercritical	-2.1%††	Not Significant
600+ Mw Subcritical	Not Significant	Not Significant
All Units	-1.8%	+1.4%

†In Table 7.1, "unit change" is defined as a one percentage point increment for sulfur and ash content. Coal quality variables are indexed to represent categories of coal quality (see Section 6.2). Figures in Table 7.1 are close approximations to the rate of change of capacity performance per change in coal quality measured as a continuous variable.

††The -2.1% figure for sulfur content is the average of the sulfur content coefficients in Equation 6.12 (−1.94%) and Equation 6.13 (−2.34%) for large supercritical units.

related to capacity performance only for 400-599 Mw units. Moisture content was not tested in the trend equations because of its strong inverse relationship to heat content.

Sulfur content reduction of coal capaicty performance results from corrosion by sulfates and sulfur trioxide which are formed from combustion of sulfur-bearing coal. For example, boiler tube leaks due to high-sulfur (4%) coal corrosion have been cited as a cause of capacity factor reductions at Illinois Power's Baldwin 1 (623-Mw).[2] Similarly, air heater corrosion caused a 12-week outage at TVA's Paradise 3 (1150-Mw) in early 1976 and is attributed to use of 4%-sulfur coal.[3] Some sulfates have low melting points and tend to deposit on boiler tubes. This attracts fly ash particles until thick slag formations develop which inhibit heat transfer and damage boiler tube surfaces.

Increases in capacity performance related to coal ash content for sub-600 Mw units run counter to the notion that the heavier concentration of impurities in high-ash coals creates greater equipment wear. However, the relationship between ash content and slagging potential is complex. The tendency for fly ash to deposit as slag on boiler tubes and other surfaces is not a direct function of coal ash content, but is related to the ash softening temperature, or "fusibility." If fly ash is subjected to temperatures over the threshold at which it softens, "it will form a fused mass that is tightly bonded and hard to remove."[4]

Many coals with high ash content also have high ash fusion temperatures[5] so that high ash content could be correlated with a tendency to avoid slagging, depending upon the grade used. Although some sub-bituminous (Western) coals have a high ash content and a low fusion temperature, the ash deposits that result are generally non-corrosive. Moreover, they are lightweight and loosely bonded and so are more easily removed by conventional soot-blowing.

The formation rate of slagging deposits on boiler surfaces is also influenced by design and operating practices. Boiler size, burner placement, firing temperatures, and soot-blowing capability can significantly affect slag formation.[6] Failure to match boiler characteristics to the coal being burned can significantly affect capacity performance. Failure to make such a matching is frequently cited by utility engineers as a major cause of reduced capacity factors for coal-fired plants.[7] Mismatches can result from (i) incorrect specification of coal quality characteristics when purchased, especially at mine-mouth plants which are "locked in" to a single coal source; and (ii) changes in the coal type delivered due to air pollution or mine safety requirements that affect coal allowed or mining practices.*

Section 7.4: Unit Size

The trend equations conclusively show capacity performance to be inversely related to unit size for coal-fired units (see Chapter 6). For supercritical units, capacity performance has declined 1.1% per 100 Mw increase. Subcritical units declined 5.6% per 100 Mw for units under 400 Mw and 2.2% for those above.

Lower average capacity factor of large nuclear units is based on their use of (i) more fixed size components; or (ii) a set number of fixed size components in more demanding service conditions; or (iii) a set number of large components which are inherently more difficult to assemble and repair (see Section 5.2). The same considerations apply to large coal-fired units. Effect of unit size on coal capacity performance above 400 Mw is somewhat smaller (about 1%-2% per 100 Mw) than the effect for comparably-sized nuclear units (about 3.3% per 100 Mw). The difference may be due to the greater likelihood of nuclear unit shut-downs caused by partial equipment failure—a result of tighter regulatory constraints applied to nuclear plants.

Opinion is mixed among power engineers on the effect of unit size on capacity performance and appears to derive from the experience of an engineer's own utility rather than from comprehensive study. TVA, which operates large units of subcritical and supercritical design, reports outages are longer at large units because of repair magnitude inherent in their size. Recent repair of air heater corrosion damage at TVA's Paradise 3 (1150-Mw) took 12 weeks, considerably longer than would have been required at a smaller unit. This was simply a result of the larger area to be repaired. Larger repair crews are not necessarily a solution: "We try to put in as many men as possible for a bigger repair," says TVA, "but you can load a job with just so many men until they fall over each other. You reach a point of diminishing returns with manpower."[8]

TVA also notes that its large units require use of more complex automatic control systems to regulate the flow of boiler feedwater, air, and coal.[9] Similarly, Central Illinois Public Service contends that larger unit size results in more frequent forced outages due to more complex controls required, and because more equipment and material means more chance for failures that can cause shutdown: "From just the larger number of feet of tubing in the boiler, your chances of having a leak are greater, so you'll have to come down more often."[10]

*Statistical analysis of capacity performance impact of changes in characteristics of coal burned by power plants is beyond the scope of this study.

TABLE 7.2: AVERAGE CAPACITY FACTOR FOR AEP LARGE UNITS†

Unit Size	Average Capacity Factor	Number of Units	Number of Unit-Years
615Mw	68.5%	3	17
816Mw	64.1%	5	11
1300 Mw	62.4%	2	3

†The 1300-Mw unit-years are Amos 3 (1974, 1975) and Gavin 1 (1975). These units were installed too late for inclusion in the CEP data base. Net generation data was obtained from AEP Form 1 reports to FPC, Schedule 432b. Note that Table 7.2 gives capacity factor rather than capacity performance data.

American Electric Power, however, notes that performance of its units is consistent regardless of unit size. This is borne out by comparison of AEP average capacity factors for its three large unit sizes—615-Mw, 816-Mw, and 1300-Mw—in Table 7.2.

Slightly higher average capacity factor of smaller AEP units is consistent with their greater average unit ages, based on the average maturation of large supercritical units (see Equations 6.12 and 6.13).* The three unit sizes have been effectively operating at an equivalent capacity factor level and so, "Reliability is a question of design, not size," says AEP.[11]

AEP attributes the high reliability of its 1300-Mw units to its use of a "series concept," by which they have scaled up to higher unit sizes from designs which have been tested and proven on each former largest series. Moreover, AEP uses conservative design specifications that incorporate large furnaces, wide spacing between boiler tubes, gas recirculation and tempering, and spare pulverizer capability to ensure high reliability.[12] AEP's good results with very large coal units appear to be unique in the electric utilities industry among very large coal-fired or nuclear unit users.

Section 7.5: Boiler Type: Subcritical vs. Supercritical

Supercritical coal units over 400 Mw capacity operate at higher capacity performance levels than subcritical units of equivalent capacity and age, burning comparable quality coal (see Chapter 6). The differential between boiler types, on a 10-year average levelized basis, is 3.8% (400-599 Mw) and 4%-10% (600+ Mw). Similarly, supercritical units have scaled better than subcritical units. They average reductions of only 1.1% per 100 Mw increment, compared to 4.3% for subcritical units (5.6% below 400 Mw, 2.2% above).

This contradicts the widespread impression among power engineers that supercritical design is less reliable than subcritical.** It is probable that the

*There is no adjustment for somewhat higher sulfur content in AEP's 615-Mw units. But it would change anticipated capacity factors by no more than several percentage points.

**For example, *Power Engineering* noted in a 1971 roundtable interview that: "Overriding demands are for high reliability, high availability, and fast delivery and startup. Supercritical pressure boilers have been given bad marks in all of these areas."[13]

TABLE 7.3: CAPACITY PERFORMANCE OF COAL-FIRED UNITS AFTER AGE 9†

	Capacity Performance/Number of Units			
Age	**100-199 Mw**	**200-299 Mw**	**300-399 Mw**	**400-599 Mw**
10	82.6%/31	77.6%/19	70.3%/9	60.0%/4
11	79.7%/24	78.1%/17	69.0%/10††	59.6%/3
12	79.6%/15	72.4%/10	64.3%/8	45.9%/3
13	78.1%/11	75.6%/5	78.9%/3	76.1%/2

†All units in Table 7.3 were 1960-63 installations. This sample selection reduces potential vintage effect which could otherwise mask the negative effect of unit age on capacity performance.

††Ten units of 300-399 Mw operated at Age 11. But only nine units operated at Age 10, due to one's exclusion from the coal data base because the percent of fuel input provided by coal fell below 60% in one year.

perception of supercritical units as less reliable than subcritical results from the larger average capacity of supercritical units, not from the use of a higher pressure cycle.

Supercritical units are inherently more complex in design and operation than subcritical. The following are some added features required to achieve high-pressure boiler operation:

1. A demineralizer system to maintain high feedwater purity. This is necessary as impurities entering with feedwater are more likely to deposit in the boiler or other steam-generating equipment in supercritical units.

2. More complex control and bypass systems to maintain correct temperatures and pressures in the boiler and the turbine.

3. More complex systems required for unit start-up due to the "once-through" flow of steam, in contrast to the "drum-type" design of subcritical boilers.

Conversely, greater heat value per unit of steam in supercritical units permits initial turbine stages to be sized smaller than they would be on an equal-capability subcritical unit. This should reduce turbine-related outages. Moreover, supercritical steam is less prone to condense in the turbine and so should result in reduced blade corrosion, a major source of turbine outages. However, a comparative analysis of outages at subcritical and supercritical units has not been performed.*

Section 7.6: Unit Age

Capacity performance decreases by slightly over 1% per year for each additional year of operation by coal-fired units under 600 Mw (1.0%/100-399 Mw, 1.3%/400-599 Mw) (see Chapter 6). This is consistent with the prevailing view that general plant wear-and-tear reduces operating availability at a gradual but steady rate.

*One question that may affect the subcritical/supercritical issue is whether higher capacity performance of supercritical units, statistically based on 400+ Mw units, is related to supercritical cycles being specifically designed for large units. Introduction of supercritical units coincided with the trend toward large unit sizes, whereas subcritical design was developed for small units.

Capacity performance typically peaks at Age 2 for units under 400 Mw, indicating a break-in period of about a year. Units of 400-599 Mw had no such break-in period, however.*

Average decline in capacity performance for coal-fired units after Age 10 is also on the order of 1% per year, as shown in Table 7.3.

The slight average decline in capacity performance for coal units after Age 10 contrasts with the rapid drop in capacity factors for the three 200-Mw prototype nuclear units after their tenth year of operation (see Section 5.3).

For large coal units (600+ Mw), the capacity performance of individual units has increased over time. Average increase for both subcritical and supercritical large units has been 4.4% from Age 1 to Age 2, and another 3.3% to Age 6 (see Section 6.5). This is about one-sixth less than the maturation tendency of nuclear PWR's. The limited data beyond Age 6 for 600+ Mw subcritical units shows continuing increases in capacity performance (see Table 6.13). Large supercritical units have not operated past Age 6 in the CEP data base.

Section 7.7: Company Operation

Because almost two-third of the large coal-fired unit-years in the CEP data base are accounted for by only four operating companies, trend equations tested for potential capacity performance effect of design and operating procedures by these companies (see Chapter 6). Supercritical units operated by AEP and Penelec are 9.3 percentage points higher and 10.5% lower, respectively, than comparable units of other utilities burning comparable quality coal. Subcritical units of TVA and Commonwealth Edison are 9.6% higher and 9.9% lower, respectively, than other utilities with comparable units.** Error coefficients for these terms are small, indicating the coefficients have a high degree of statistical confidence.

There are perils inherent in comparing capacity factor data for different companies. The extent of load-following may vary significantly; or companies may assign different capacity ratings to comparable units. Even so, the sizeable difference in capacity performance for the four utilities is significant.

Penelec operates four 936-Mw units and two 660-Mw units (and others under 600Mw), all at mine-mouth stations in Western Pennsylvania. At installation their 936-Mw units were the largest supercriticals ever built in the U.S.*** They were apparently an over-ambitious undertaking by a utility which at that time, "had little in-house engineering capability, a very small operating staff, maintenance capability oriented to smaller size plants, and

*"XA," a "dummy" variable defined to equal one when AGE=1 and zero when AGE≠1, has a value of -8.0 for 100-399 Mw units, but was not statistically significant for 400-599 Mw units.

**All four utilities operate only subcritical or supercritical units, about 27 each, except TVA. It accounts for nine supercritical unit-years in addition to 20 subcritical unit-years (see Tables 6.11 and 6.14). TVA operation was not a significant variable in the regression equations for supercritical units.

***The 936-Mw units entered commercial service in 1967, 1968, 1970, and 1971. They have since been exceeded in capacity by TVA's Paradise 3 (1150-Mw) and the three 1300-Mw units of AEP.

minimal industry or vendor experience . . . ," according to a report prepared for FEA.[14]

For Keystone 1, the first large Penelec unit, the architect-engineer attempted to design a highly efficient plant using "all of the latest design improvements at a very low cost."[15] "Design margins"—the range of operating capabilities specified in component design—were often inadequate. For example, the furnace is sized too small. The result is inadequate heat transfer causing boiler hot spots that lead to serious slagging problems. These are then exacerbated by inadequate soot blower capacity. Moreover, the mine-mouth coal, which did not meet design specifications at start-up, has since deteriorated in heat content and ash fusibility because of mine safety regulations causing inclusion of mine debris with coal. This has also encouraged slagging and further capacity reductions from inadequate coal pulverizer capability.

These problems and resultant poor capacity factors are mirrored at Penelec's five other large coal units. Problems originated from failure to design "adequate margins for operability and maintainability." These fundamental design problems have been intensified by deteriorating coal, according to the FEA consultant's report.[16]

AEP, on the other hand, employs a conservative design philosophy reflected in use of large boilers and heavier boiler tubing to prevent leaks. AEP also designs its units with more margins for error (*e.g.*, additional coal pulverizer capacity) to the point of maintaining a large supply of spare parts to reduce outage durations.[17] These practices add to capital costs, but are considered to be ultimately cost-effective. Design philosophy and AEP's large in-house engineering staff give the company a strong reputation for reliability.

Unfortunately, only further analysis can explain the disparate performance of TVA's and Commonwealth Edison's large units. Higher capacity factors of TVA's large subcritical units, Paradise 1 and 2, may be partially due to their heavy base-load operation under the utility's "economic dispatch" system. The coal burned at Paradise is the lowest-cost on the TVA system. Since the CEP capacity performance model estimates large coal units load-follow at 7.2% capacity factor per year, and since Paradise 1 and 2 probably undergo some load-following, it cannot account for all of the 9.6% capacity performance superiority of TVA units over other subcritical units.

No explanation is readily available for the sub-par capacity performance of Comm Ed's subcritical coal-fired units. Load-following due to Comm Ed's operation of nuclear units with lower marginal running costs does not appear to be an explanation since 18 of Comm Ed's 27 large subcritical unit-years in the CEP data base occurred prior to 1972, the first year that Comm Ed generated a substantial amount of electricity from nuclear units. Moreover, the capacity factors of the Comm Ed coal units did not decline in 1972 and 1973, indicating that load-following probably did not increase significantly in those years (the CEP data base for coal ends with 1973). Similarly, coal quality switching, which has occurred at numerous Comm Ed plants, has not taken place at the company's large units. Some capacity factor reductions have resulted from electrostatic precipitator upgradings, but not enough to account for the 9.9% lower Comm Ed capacity factors indicated in Equation 6.16.*

*By letter dated July 8, 1976, Commonwealth declined to respond to a CEP request for an explanation of the low capacity factors.

REFERENCES

1. Telephone communication with Gilbert Moorman, Central Illinois Public Service Co., July 9, 1976.

2. Nuclear Services Corp., "Productivity Improvement Action Plan," report to FEA, November 28, 1975, p. 34.

3. Telephone communication with Al Spencer, TVA, July 14, 1976.

4. John Graham, "Combustion Optimization," in *Electrical World,* June 15, 1976.

5. *Ibid.*

6. R.J. Gray and G.F. Moore, Babcock & Wilcox Co., "Burning the Sub-Bituminous Coals of Montana and Wyoming in Large Utility Boilers," paper presented at the Winter Annual Meeting of The Fuel Division of The American Society of Mechanical Engineers, November 17-22, 1974.

7. See, for example, Trident Engineering Associates, "Power Plant Productivity," report to FEA, November 17, 1975, *passim.*

8. Telephone communication with Al Spencer, TVA, July 14, 1976.

9. E.F. Thomas, J.G. Holmes, Jr. and C.C. Schonhoff, TVA, Operating Experience With Large Supercritical Units," *Combustion,* April 1971, pp. 13-14.

10. Telephone interview with Gilbert Moorman, Central Illinois Public Service Co., July 9, 1976. Moorman was citing his utility's 617 Mw Coffeen 2 unit which, despite its lower reliability as compared to smaller units, has the lowest total generating costs because of economies of scale in the capital cost which apparently outweigh the lower reliability. Tradeoffs between capital cost and capacity performance for different unit sizes are discussed in Chapter 9.

11. Telephone communication with Gregory Vassell, Vice-President, System Planning, AEP, July 9, 1976.

12. John Tillinghast, AEP Vice Chairman, Engineering & Construction, and John E. Dolan, AEP Sr. Executive Vice-President, Engineering & Construction, "AEP Succeeds With Large Units," *Electrical World,* August 1, 1976. The article also reports on the initial performance of AEP's third 1300-Mw unit, Gavin 2, and concludes that the shakedown problems of the three 1300-Mw units to date have been less than those of previous smaller units.

13. "Supercritical Boilers: Confusion Abounds and the Trend is Down," *Power Engineering,* April 1971, p. 32.

14. Trident Associates, "Power Plant Productivity," November 17, 1975, p. 20.

15. *Ibid.*, p. 20.

16. *Ibid.*, pp. 32-33.

17. Telephone communication with Gregory Vassell, *op. cit.*

8: The Prognosis for Coal-Fired Plant Capacity Performance

Section 8.1: Introduction and Key Findings

Coal-fired power plants have not generally been assigned the 80% target capacity factor projected for nuclear units until recently. This is in part because capacity factors of coal-fired units tend to be influenced by system load characteristics, rather than solely by mechanical reliability concerns such as those that have determined nuclear capacity factors.

Nevertheless, in light of generally high operating reliability of small coal-fired units—100-299 Mw units in the CEP data base averaged 71% capacity factor and 80% capacity performance—the lesser capacity statistics for larger, modern coal units have unquestionably been a disappointment to utilities. Coal-fired 400+ Mw units have operated at average capacity factors of 55%-60% and only a 60%-65% capacity performance rate. Coal-fired units of 400+ Mw can be projected to operate at a capacity performance of 55% to 70%, with specific values dependent upon size, boiler type, sulfur content, and use of scrubbers (see Table 6.1).

Capacity performance of coal-fired units is predicted to be higher than for nuclear. Projected capacity factors of nuclear plants from the trend equations are in the low-60's for 600-Mw nuclear units, but only 45% for 1150-Mw units (see Table 3.1). The larger is the predominant nuclear unit size being planned and constructed by U.S. utilities. As noted, however, capacity factors for future nuclear units will probably exceed those derived from the trend equations due to anticipated improvements in design, operation, and maintenance (see Chapter 5). A similar likelihood exists for coal-fired units.

The major findings for coal-fired units are as follows:

1. There is now more economic incentive for improving coal plant capacity factors then there was when coal units in the CEP data base were installed, although the incentive increase is less than that for nuclear plants. This increased incentive will likely be translated into more conservative design and construction practices to improve reliability.

2. Unit duplication and, to a lesser extent, design standardization can

improve coal-fired capacity performance, though potential is probably less than that for improvement of nuclear plants.

3. Scrubbers will reduce capacity performance of new coal-fired units below CEP capacity performance trend equation projections. The approximate reduction will be 5%, or 3-3½ percentage points for large units.

4. Since 1970, environmental and mine safety regulations have led to reduced capacity performance in many coal-fired plants by company compliance efforts that have caused changes in moisture and ash characteristics of coal burned in boilers and by required plant shutdowns to upgrade pollution control equipment. However, anticipation of regulations prior to design and construction of future plants should reduce outages related to air quality and mine safety regulation.

5. Utility action toward improved matching of boiler specifications to coal grade burned and closer adherence to coal quality specifications by coal suppliers should reduce outages in boilers and related steam-generating equipment.

Section 8.2: Incentives for Improving Coal Plant Reliability

Incentive for upgrading a unit's capacity factor is directly proportional to the difference between (i) the unit's marginal running cost (chiefly fuel cost) and (ii) the marginal running cost of the unit(s) which must be operated if the first unit is unavailable.* While this difference between the *unit's* marginal cost and the *system*'s marginal cost is anticipated to be less for coal than nuclear units, it is still greater today than during the late 1950's and the 1960's when the coal plants in the data base were being designed and constructed. This is a result of the general increase in the price of fuel-oil and natural gas relative to coal price.

Capacity factor benefits in reliable design are a trade-off against the increased capital cost which often accompanies such an upgrade. American Electric Power has consistently opted for more conservative designs for its coal units on the grounds that the trade-off is cost-effective (see Section 7.7). Similarly, Minnkota Power in North Dakota is reported to have purchased a boiler for a new unit at a cost 10% over low bid specifically in anticipation of greater reliability with the more expensive boiler.[1]

Although such practices are situation-based, it is likely that incentives for higher capacity factors will reflect in new unit design, presumably with higher capital but lower life-cycle costs. There are, moreover, some increases in reliability that may result from a feedback learning process via operating experience and technological improvements with no requisite cost addition.

Section 8.3: Duplication and Standardization

As is the case with nuclear, duplication and standardization of coal-fired power plant design can be expected to improve capacity factors without raising capital costs (see Section 5.5). As noted, *duplication* is identical unit construction at the same site, with the second unit slightly lagging the first.

*An indirect incentive is the potential to reduce overall system capacity requirements through constructing and maintaining units which are less prone to forced and scheduled outages.

Standardization is identical major equipment construction. However, the overall structure and layout and much of the electrical and piping systems of standardized units must still be geared to the site.

Coal-fired plants are not as amenable to industry-wide design standardization because of coal grade-boiler characteristic match variations. However, some standardization of non-boiler components is possible. Moreover, boiler standardization should be possible on a regional basis, since there is frequently less variation in coal characteristics within given regions. Duplication of design is certainly feasible on a station basis.

As with nuclear PWR's (see Section 5.5), the trend analysis tests large (600+ Mw) supercritical units for the relationship between capacity performance and duplicate units, deriving Equation 8.1.

(Equation 8.1)

$$\text{Capacity Performance} = 76.64 - \underset{(1.44)}{3.81} \times \text{AGED} + \underset{(2.88)}{8.73} \times \text{AEP}$$

$$- \underset{(3.05)}{11.98} \times \text{PENN} - \underset{(1.02)}{2.01} \times \text{XSUL} + \underset{(2.59)}{4.87} \times \text{DUPE*}$$

$$N = 108$$
$$\text{Adjusted } R^2 = 0.31$$

Equation 8.2 provides a 10-year levelized version of Equation 8.1, assuming 2% sulfur and operation not by AEP or Penelec:

(Equation 8.2)

$$\text{Capacity Performance} = 65.18 + 4.87 \times \text{DUPE}$$

Equation 8.2 suggests a duplicate large supercritical unit to be four percentage points higher than indicated for large supercritical units in general by Equation 6.15. A unit that was not a duplicate would be 1% lower than previously indicated, and 5% lower than a duplicate unit. Although the 4% improvement for duplicate large supercritical coal units is less than the 9% registered for duplicate nuclear PWR's (Equations 5.1 and 5.2), the coal result is based on a sample of 30 unit-years as compared to 11 for PWR duplicate units.**

*"DUPE" is defined to equal one if a unit entered commercial service roughly 6-24 months after an equal-size unit at the same site, and zero otherwise. The CEP data base includes 11 such duplicate 600+ Mw supercritical units operating for 30 unit-years. These units are identified in the CEP data base listing in Appendix A.

As with PWR's, capacity factors of "older brothers" of large supercritical coal duplicate units showed no statistically significant difference from those of non-duplicate units, thus demonstrating that higher average capacity factor of duplicate units has been due to duplicate design.

Subcritical 600+ Mw units were not tested for the effect of duplicate units since there are only four. Smaller size groups were not tested due to difficulty of identifying duplicate units for large sample spaces.

**Inclusion of seasonally adjusted nuclear capacity factor data from the first half of 1976 reduces the projected capacity factor improvement for duplicate PWR's from 8.8% to 5.6% and reduces the statistical significance of the improvement to the 93% confidence level (see Section 5.5 and Appendix D).

Section 8.4: Coal Mining and Air Quality Regulations

The greatest impact on coal capacity factors will likely come from changes in coal quality and air pollution control equipment based on federal mine safety and clean air regulations. A significant number of new coal-fired units will be required to have "scrubbers"—devices which wash flue gases with a chemical solution to eliminate sulfur dioxide. The "New Source Performance Standards" promulgated in 1971 by the federal Environmental Protection Agency, pursuant to the 1970 Clean Air Act, require that new coal-fired plants limit their sulfur dioxide emissions to the level resulting from burning approximately 0.7%-sulfur coal without sulfur dioxide controls.* Since most U.S. coal contains sulfur in excess of 0.7%, many new plants will need scrubbers to meet EPA standards.**

Operating experience of scrubbers on U.S. power plants is limited. Only two of 250 coal-fired units in the CEP data base had operated with or otherwise tested scrubbers by the end of 1973 and both were only 114 Mw capacity,*** much less than typical new unit size. Twenty-seven U.S. units had scrubbers by mid-1976, but virtually all of them were very old or very new units rather than those installed during 1960-72, the installation period covered by the coal data base.

Accordingly, the prospective impact of scrubbers on the reliability of new coal plants cannot be confidently estimated. The impact would be expected to be negative. To date some scrubbers have caused scaling on boiler tubes due to attachment of sulfur solution precipitate to boiler surfaces. Such problems should diminish with improved scrubber design and operation, but it cannot be assumed they will end. Scrubbers are complex systems adding 15%-20% to plant capital cost, while adding nothing to operating reliability and potentially reducing it.

Moreover, scrubbers will also experience outages. They will not necessarily impair the plants' mechanical reliability, but may lead to significant de-rating or outright shutdown in order to comply with air quality regulations. While the compliance consistency debate over air standards is unresolved, it is likely that utilities will not be permitted to routinely by-pass scrubbers in the event of malfunction. So de-ratings would be required in some instances. Alternatives

*The standards prescribe that emissions be limited to a rate of 1.2 lb of sulfur dioxide per million Btu of coal burned. Coal with a typical heating value of 11,667 Btu/lb will produce sulfur dioxide at such a rate if it contains 0.7% sulfur by weight.

**Alternatives to scrubbers include (1) selection of low-sulfur coal to meet the New Source standards without scrubbers; (2) coal "cleaning"—a process for mechanically removing sulfur-bearing compounds from coal which are not chemically bonded to the carbon in the coal; and (3) new combustion techniques which remove sulfur from coal as it is being burned in the boiler. The first two are currently available for some new plants. They are briefly described in testimony by CEP before the New Jersey Board of Public Utility Commissioners in June 1976.[2] The third is being tested on a pilot scale in processes such as "fluidized-bed" boilers.

Conversely, application of the "significant deterioration" provisions of the Clean Air Act or of stricter state air quality regulations may require that coal-fired units burning "conforming" coal nevertheless install scrubbers to reduce sulfur dioxide emissions to the lowest possible level.

***The two units are Cholla 1 of Arizona Public Service with scrubber startup in October 1973 and Nevada Power's Reid Gardner 2, December 1973.

are available, in theory, to continuously meet emission limitations. Scrubbers can be built in modules with spare units or supplies of low-sulfur coal can be maintained for use when scrubbers fail. But they are not necessarily cost-effective and may have reliability consequences of their own (e.g., fuel switching).

Utilities now assume that scrubbers will reduce capacity factor by 5 percent. For example, Northeast Utilities has estimated that a hypothetical new coal-fired unit rated at 78% capacity factor would operate at 74% ($0.78 \times 0.95 = 0.74$) if a scrubber were added.[3] Application of this estimate would reduce the capacity factor of a coal unit assumed to operate at a 60%-70% capacity factor without scrubbers by 3 to 3½ percentage points.*

Use of scrubbers is not reflected in the CEP coal data base, so new plants that use them will have anticipated capacity performances *below* levels suggested by the trend equations (see Chapter 6). Conversely, federal regulations to strengthen mine safety and reduce emissions at existing plants are already largely accounted for in the data base. Their anticipation in future plant design and coal purchasing practices should lead to more stable operating reliability.

Both the 1969 Coal Mine Safety and Health Act and the 1970 Clean Air Act have caused switching in the characteristics of the coal burned at many U.S. power plants. Spraying to control dust in mines and the required removal of additional mined material from mines under the 1969 Mine Safety Act have added to coal moisture and ash content at some coal-fired stations. Increased moisture in coal can impair operation of coal pulverizers, change temperature conditions in and around the boiler from design conditions, and reduce effective heat transfer. This results in less net generation per unit of coal and therefore reduces capacity factor directly. Increased ash content reduces pulverizer efficiency and generally enhances boiler slagging, notwithstanding the positive correlation between capacity performance and ash content for sub-600 Mw units. This is apparently a function of ash type, not ash content.

At TVA's 950-Mw Bull Run and 550-Mw Widows Creek 8 units, the ash content is about one-third higher than original design specifications, greatly increasing coal pulverizer outages and maintenance. Pulverizer systems at both units were originally designed to provide sufficient coal to permit operation at full capacity with only 9 out of 10 pulverizers. Presently, however, use of all ten fails to provide sufficient coal due to the added ash content and consequent lower heating value per unit of coal fed to the furnace. De-ratings resulting from added pulverizer outages and insufficient capacity account for 2% capacity factor loss at Bull Run and 3½% at Widows Creek 8 in recent years.

Similarly, Pennsylvania Electric's Keystone 1 unit has suffered insufficient pulverizer capacity caused by increased coal moisture and ash content. This is a result of newly required mine spraying and the mix of debris with the coal due to work area cleanliness regulations that demand non-coal material's removal from the mine.[5]

*Use of higher-sulfur coal would thus appear to carry a double penalty for coal unit reliability: (1) a capacity factor reduction attributable to scrubbers, and (2) a reduction of about two percentage points per 1% increment in sulfur content, due to boiler corrosion from sulfur-bearing gases and compounds created in combustion (see Table 7.1). This may increase the inducement for utilities to seek out low-sulfur coal and/or coal cleaning techniques.

TABLE 8.1: COAL CAPACITY PERFORMANCE BY YEAR OF OPERATION

Year	Average Mw Capacity	Average Capacity Performance	Year	Average Mw Capacity	Average Capacity Performance
1961	226	79.6%	*1968*	302	78.3%
1962	255	80.2%	*1969*	317	76.2%
1963	243	81.0%	*1970*	334	71.6%
1964	252	82.0%	*1971*	352	70.0%
1965	250	82.3%	*1972*	382	69.1%
1066	264	79.9%	*1973*	400	69.3%
1967	280	79.9%			

Moreover, many plants have been switched to lower-sulfur coal with different ash fusibility characteristics to meet the sulfur dioxide emission limitations prescribed in the "State Implementation Plans" under the Clean Air Act. Coal switching is almost universally regarded by power engineers as a significant factor impairing reliability (see Section 7.3). "The key isn't the quality of the coal as much as it is whether that's the quality the boiler was designed for," says an engineer on the Federal Energy Administration's Power Plant Productivity Task Force.[6] Each of the consultant reports prepared for FEA cites instances in which considerable capacity reductions have been caused by coal quality changes.

Switching to lower-sulfur coal has reduced instances of sulfur-caused corrosion at some units, but it also has increased instances of boiler-slagging. This is due to the lower ash-softening temperature of low-sulfur coal, especially Western sub-bituminous. Moreover, the efficiency of conventional electrostatic precipitators is reduced with lower-sulfur coal. This has led to long outages to upgrade precipitators at a number of units switched to low-sulfur coal.[7]

Strengthened air quality regulations have also required precipitator backfitting at many units. For example, Duke Power's highly reliable 650-Mw Marshall 4 unit (76.8% capacity factor for 1971-73) lost 800 hours of operation in 1972 (equivalent to 9% capacity factor) to install more precipitators to meet air quality regulations.[8] Similar outages were experienced by most U.S. utilities during the early 1970's.

No estimate is available of the total capacity factor reduction at coal-fired units due to mine safety and air quality regulations. However, Table 8.1 suggests a "break point" in 1969-1970 after which capacity performance fell significantly. The 1966-69 downward capacity performance trend is statistically attributable to the increase in average unit size and perhaps to differences in capacity definitions associated with later vintage units (see Section 7.2). However, the sharp capacity performance drop in 1970 coincides with implementation of the 1969 Mine Safety Act, as well as precipitator upgradings and fuel switching to limit emissions in response to federal and state regulation and pressure by citizens' groups.

New plants, however, can be designed to accommodate and sometimes anticipate such regulations. Many coal-fired units are being equipped with electrostatic precipitators designed to remove 99.5% of the particulates from stack gases. This is as compared to typical 75%-95% efficiencies at units upgraded in the early 1970's.[9]

It is conceivable that future environmental regulations may require that even 99.5% efficient precipitators be upgraded. But it is more likely that regulation, having cleaned up the more egregious power plant polluters, will focus on controlling growth in energy demand and in limiting emissions at new power facilities.* The outlook suggests less fuel switching and less backfitting of improved pollution control systems onto new plants.

Similarly, the effect of poorer coal quality that results from compliance with mine safety regulations can be offset by power plants designed with more conservative margins (e.g., spare pulverizers, more soot blowers, thicker boiler tubes, larger boilers) that can efficiently handle poorer coal or variable coal characteristics. Such effects can also be minimized by closer adherence to contract specifications of coal suppliers. These alternatives will probably add to plant capital or fuel costs, but they should be cost effective by improving capacity factor.

Finally, utilities can work more closely with power plant vendors in specifying characteristics of the coal to be burned. A high-ranking engineer for a leading boiler vendor suggests such utility cooperation has been limited,

> "[Only] about a third of the customers make an honest effort to give us a true range of ash analyses to show what the fusions are . . . to help the designer of the furnace. The other two-thirds still give us something out of a book somewhere that doesn't tell us much of anything. It may cover the whole world's range from Australian brown coal to anthracite."[10]

While vendors and utilities often differ in identifying causes of operating reliability problems, power engineers are increasingly identifying the proper matching of boiler specifications to coal grade as a major consideration in plant design. Changes in furnace size, burner location and direction, temperature control, and soot blowing equipment are being initiated in an effort to reduce boiler outages at new plants.

*Federal EPA, for example, has been more stringent in enforcing the New Source Performance Standards which will lead to scrubbers at many new plants, than in enforcing the State Implementation Plans which would require installation of scrubbers at some existing facilities.

It can be anticipated that new coal-fired plants will operate at higher capacity factors than suggested by the trend equations based on the CEP data base. This is due to the positive influence of unit duplication and standardization, improved matching of boiler design to coal, design anticipation of mine safety and air quality regulations, and general learning from operating experience feedback. However, scrubbers on many new units will reduce the magnitude of potential capacity factor improvements. And, as with nuclear units, the magnitude of any such improvement is speculative.

REFERENCES

1. Interview with Bob Nelson, engineer, Office of Fuels & Facilities, Federal Energy Administration, June 18, 1976.

2. Charles Komanoff and Raymond E. Gerson, "Economic Analysis of the Nuclear Expansion Plan of Public Service Electric & Gas Co.," testimony prepared for the New Jersey Department of Public Advocate before the New Jersey Board of Public Utility Commissioners, Docket No. 761-8, June 7, 1976.

3. Testimony of Walter Schultheis, Exhibit 6, Northeast Utilities, before the Connecticut Public Utilities Control Authority, February 1976.

4. Based on the estimates by Mechanics Research Inc. that pulverizer problems due to high ash-content coal had caused 35 of the 79 de-ratings at Bull Run and 46 out of 147 de-ratings at Widows Creek 8 in the period studied, and that de-ratings as a whole had reduced capacity factors by 4.7% at Bull Run and 11.2% at Widows Creek 8. From MRI report to FEA, "Study on Reliability of Large Coal-Fired & Nuclear Power Plants," December 15, 1975, p. 80, Volume I; p. 71, Volume II.

5. Trident Engineering Associates, "Power Plant Productivity: An Assessment of Outages at Two Coal-Fired and Two Nuclear Plants," report prepared for FEA, November 17, 1975, p. G-5.

6. Interview with Bob Nelson, Office of Fuels & Facilities, FEA, June 18, 1976.

7. See, for example, Council on Economic Priorities, *The Price of Power,* 1972, Chapter C, for an account of Commonwealth Edison's precipitator upgrading and fuel switching in the early 1970's.

8. Trident Report, *op. cit.*, p. K-9.

9. *The Price of Power, op. cit., passim.*

10. William D. Stevens, Vice President & General Manager of the Equipment Division, Foster Wheeler Co., in "Supercritical Boilers: Confusion Abounds and the Trend is Down," roundtable interview in *Power Engineering,* April 1971, p. 37.

9: Capacity Performance and Optimal Power Plant Size

Section 9.1: Introduction and Summary

From 1882 through the present, electric utilities in the U.S. and elsewhere have consistently increased the unit size (megawatt capacity) of new power generating units. Several factors have led to this trend, foremost being the reduction in the per-unit capital cost of new units through economies of scale inherent in construction of larger units.*

However, installation of large generating units has also resulted in diseconomies of scale. Most significant is the reduced average operating reliability of larger units, both nuclear and coal-fired. This adds to utility costs at the system level at which additional capacity must be installed to ensure a given reliability of electric service, and at the plant level at which the total per-kwh generating cost is increased by lower capacity factors of larger units.**

Following are our major findings on unit size as it relates to (i) capital cost and (ii) capacity performance.

1. For nuclear and coal-fired plants, capital costs generally decrease for larger unit sizes, but capacity factors also decrease. Based on the capacity factor trend equations, unit sizes at which capital charges per kwh generated are minimized are: 600 Mw for PWR and BWR nuclear units, and 650 Mw for subcritical and supercritical coal units. The per-kwh capital charges vary for each plant type by no more than several percentage points within 200 Mw of these unit sizes.

*Other factors leading to larger unit sizes have been their historical lower operating and maintenance costs and their improved efficiency of conversion of fuel to electricity ("heat rate"). However, further improvements are apparently not occurring today with larger plant sizes (see Section 9.5).

**Use of larger units has tended to increase utility capacity requirements since (i) larger units generally have higher outage rates than smaller units, thereby requiring greater reserve capacity, and (ii) loss of a large unit has a greater impact on system capacity so that, probabilistically, a system based on large units tends to be less reliable than a system of equal size based on small units of equal reliability, and therefore requires greater reserve capacity. The latter problem is mitigated, however, for very large utility systems.

2. Optimum size of new power plants is generally less than the minimum kwh charge unit sizes. Smaller units enhance unit duplication or siting flexibility, reduce capacity reserve margins, and cut costs of fuel, operating, and maintenance. The approximate optimal unit size ranges are 350-700 Mw for nuclear units and 350-800 Mw for coal.

3. The average capacity of nuclear units to be installed between 1975-84 is 1064 Mw. The typical nuclear unit being planned is 1150 Mw. Based on the capacity factor trend equations, the per-kwh capital charges of 1150-Mw nuclear units will exceed those of 600-Mw nuclear units by 17% for BWR's and 13% for PWR's, despite capital economies of scale of larger units. Thus, if the inverse relation between nuclear unit size and capacity factor persists, the very large nuclear units will be considerably less economical than the smaller (450-800 Mw) which comprise most of U.S. nuclear plant experience to date.

4. The average capacity of new fossil units being installed by U.S. utilities during 1975-84 is 457 Mw. The typical new fossil unit planned is in the 500-600 Mw range. Most coal-fired units being added by U.S. utilities are within the size range in which the optimal combination of low capital cost and high operating reliability is achieved.

Section 9.2: Determination of Per-Unit Generating Cost

A power plant's "annualized generating cost" is the sum of its capital-related costs, fuel costs, and operating and maintenance costs over the course of a year. "Annualized capital costs" are the product of the "capital cost" (construction cost) of the power plant and the "annual fixed charge rate" which is a percentage representing the annual cost of stocks and bonds that the utility issues to finance construction. The fixed charge rate also includes property taxes and other expenses proportional to capital costs. Currently the annual fixed charge rate is 15%-17% for most investor-owned utilities.

These annualized capital costs are constant once the power plant is built. Capital costs thus represent fixed costs which do not vary with the amount of electricity produced by the plant. Fuel costs, on the other hand, are proportional to the rate at which the plant produces electricity* and are measured on a per-kwh basis. Operating and maintenance (O&M) costs are generally measured on an annual basis, although some are proportional to the rate of plant operation (see Section 9.5).

Utilities attempt to minimize system-wide generating costs, per kwh of electricity produced, by economical construction of new facilities and by reduced running costs. This objective generally applies on a per-plant, or per-kwh, basis as well.** A plant's per-kwh fuel costs are only moderately sensitive to the rate of generation, and O&M costs are relatively small. Accordingly, minimizing the per-kwh cost of generating electricity usually entails

*This is not strictly true for nuclear units (see Section 9.5).

**There are exceptions to this rule, the most significant being that a unit's kwh production should not be increased at the expense of reduced kwh production by another unit with lower incremental running costs. This is, of course, the constraint which determines load-following through economy dispatch.

Figure 9.1

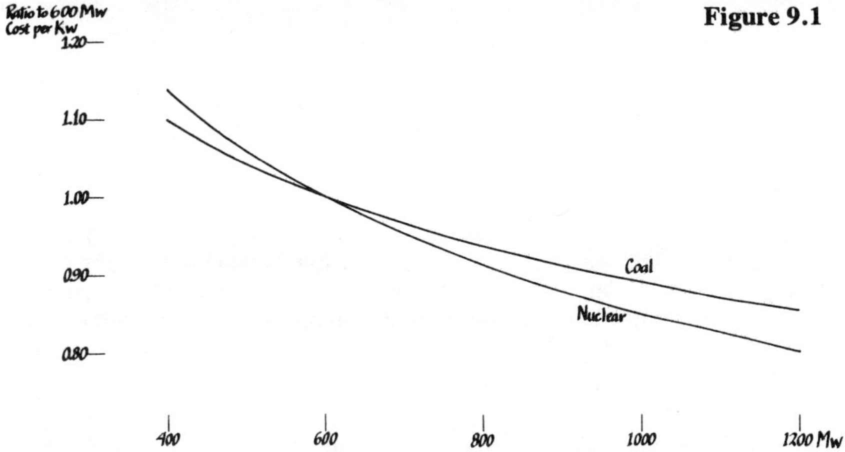

TABLE 9.1: RATIO OF PER-KW CAPITAL COST TO COST OF HYPOTHETICAL 600-MW UNITS

Unit Size	Nuclear	Coal
400 Mw	1.14	1.10
600 Mw	1.00	1.00
800 Mw	0.91	0.94
1000 Mw	0.85	0.89
1200 Mw	0.80	0.85

Note: The nuclear and coal columns give only the capital economies of scale for each plant type separately and do not describe relative nuclear-to-coal capital costs.

Costs for the ERDA study are undifferentiated by reactor type (PWR vs. BWR) or boiler type (subcritical vs. supercritical). Coal plant costs assume scrubbers.

minimizing the plant's annual capital cost per kwh's produced. This quantity is measured by Equation 9.1.

(Equation 9.1)

$$\text{Capital Cost per kwh} = \frac{\text{Capital Cost per kw} \times \text{Annual Fixed Charge Rate}}{\text{kw-hrs per kw}}$$

The expression "kw-hrs per kw" in the denominator on the right-hand side of Equation 9.1 is equal to the plant's capacity factor, multiplied by the number of hours in the year as in Equation 9.2.

(Equation 9.2)

$$\text{Capital Cost per kwh} = \frac{\text{Capital Cost per kw} \times \text{AFCR}}{\text{Capacity Factor} \times \text{hrs/yr}}$$

This may also be expressed as Equation 9.3, since both the annual fixed charge rate and the number of hours in the year are constant.

(Equation 9.3)

$$\text{Capital Cost per kwh} = \text{Constant} \times \frac{\text{Capital Cost per kw}}{\text{Capacity Factor}}$$

Equations 9.3 demonstrates that the per-kwh capital cost of a specific plant type depends on the ratio of the plant's per-kw capital cost to its capacity factor. Presently, both capital cost and capacity factor decline with increased plant size. Thus, minimization of the per-kwh capital cost and, as a first approximation, of the plant's overall per-kwh generating costs involves selecting the plant size at which the ratio of capital cost to capacity factor is least.

Section 9.3: Economies of Scale in Capital Costs of Nuclear and Coal-Fired Generating Units

The capital cost economies of scale inherent in the construction of large nuclear and coal unit sizes may be represented by Equation 9.4.

(Equation 9.4)

$$\frac{\text{Per-kw Capital Cost of Unit of Y Megawatts}}{\text{Per-kw Capital Cost of Unit of X Megawatts}} = \left(\frac{X}{Y}\right)^k$$

Here k is a number greater than 0 and less than 1.* The greater the value of k, the greater are the capital economies of scale for large units.

The value of k is approximately 0.32 for nuclear units and 0.23 for coal-fired units, on the basis of a detailed study of nuclear and coal plant capital costs being conducted by ERDA.[1] While the capital costs for future generating units in the ERDA study are lower than current estimates of future costs, the relative capital costs of different unit sizes and types in the study appear to accurately reflect current estimates.

Using 600 Mw as a reference point, substitution of $k = 0.32$ for nuclear units and $k = 0.23$ for coal yields the capital cost economies of scale shown in Table 9.1 and Figure 9.1.

Table 9.1 and Figure 9.1 show that nuclear units benefit from greater capital economies of scale than coal-fired units. For example, to achieve a 15% reduction in per-unit capital cost compared to the cost of a 600-Mw unit, a

*To grasp the way in which Equation 9.4 works, consider a case where Y = 1000 Mw and X = 500 Mw, so that $X/Y = 0.5$. If k is determined to be equal to 1, then $(X/Y)^k = 0.5^1$, or 0.5, so that the 1000-Mw unit would have a per-unit capital cost half that of the 500-Mw unit. This is obviously an extreme case of economies of scale since the two units would then have the same total dollar capital cost.

At the other extreme, if k equals zero, then $(X/Y)^k = 0.5^0$, or 1, so that the two units have the same per-kw capital cost—*i.e.*, there are no economies of scale. In practice, k falls between 0 and 1.

coal-fired unit must be sized as 1200 Mw, whereas nuclear units achieve the same cost reduction at a size of 1000 Mw.* This would account in part for greater movement by utilities to larger unit sizes for nuclear as compared to coal units.**

Table 9.1 and Figure 9.1 also indicate that capital economies of scale tend to level off for constant increments in unit size.*** For example, construction of an 800-Mw rather than a 600-Mw nuclear unit reduces per-kw capital costs by 9%, but a further increase to 1000 Mw (200 Mw increment) reduces capital costs by only 6½%. Benefits of increased unit size constantly taper, suggesting that the size penalty upon nuclear and coal capacity factors must be reduced for very large unit sizes if they are to be economical (see Chapters 3 and 6).

Section 9.4: Minimization of Nuclear and Coal Capital Charges per kwh as a Function of Unit Size

Equation 9.3 indicates that reduction of annual capital charges per kwh for a given plant type (*i.e.*, nuclear or coal) is achieved by lowering the ratio of per-kw capital cost to capacity factor. For nuclear units, the variation in per-kw capital cost as a function of unit size is given by Equation 9.4, with $k = 0.32$. The variation in capacity factor for nuclear units as a function of unit size, based on the CEP data base, is given by Equation 3.2 for PWR's and Equation 3.3 for BWR's.

*(Equation 3.2****)*

$$\text{Capacity Factor} = 86.38 - 3.368(\text{Mw}/100)$$

*(Equation 3.3****)*

$$\text{Capacity Factor} = 79.30 - 3.287(\text{Mw}/100)$$

Equations 3.2 and 3.3 may each be combined with Equation 9.4 to develop Table 9.2 of nuclear per-kwh capital charges as a function of unit size, using the capital cost and capacity factor of a 600-Mw nuclear unit as a reference point.*****

*The difference in scale economies for nuclear and coal is primarily due to the different scale economies for the nuclear steam supply system (NSSS) and for the boiler plant in a coal unit. The ERDA study estimates "k" (the scaling exponent in Equation 9.4) to be 0.4 for an NSSS, but only 0.1—virtually no economy of scale—for fossil boilers.[2]

**Other factors appear to be (i) the greater willingess of utilities to carry excess nuclear capacity resulting from large units, due to the lower fuel costs of nuclear units and (ii) the fact that smaller utilities, which do not require large units of either type, tend to opt more frequently for coal. That is, utilities going nuclear tend to be able to afford large nuclear units, while the same is not always true for coal-oriented utilities. However, utilities installing or planning both nuclear and coal-fired plants in recent years have invariably chosen a larger unit size for the nuclear plant.

***From the form of Equation 9.4, equal cost reductions are achieved through equal increases in the ratios of unit sizes rather than equal increases in the differences.

****Equation 3.2 is a ten-year levelized average capacity factor equation drawn from Equation 3.1, in which PWR capacity factor varies with unit age. Since age was not a statistically significant variable for BWR capacity factor, Equation 3.3 for BWR's already represents a levelized average.

*****The unit size for which the ratio of per-kw capital cost to capacity factor is minimized is 622 Mw for PWR's and 585 Mw for BWR's, based on Equations 3.2, 3.3, and 9.4. Since the slope of the relationship between unit size and per-kwh capital charges is very flat around 600 Mw, that unit size may be chosen as a base point for Table 9.2.

TABLE 9.2: RATIO OF NUCLEAR PER-KWH CAPITAL CHARGES TO CAPITAL CHARGES OF A HYPOTHETICAL 600-MW NUCLEAR UNIT

Unit Size	PWR's	BWR's
400 Mw	1.03	1.03
500 Mw	1.01	1.00
600 Mw	1.00	1.00
700 Mw	1.00	1.01
850 Mw	1.03	1.04
1000 Mw	1.07	1.09
1150 Mw	1.13	1.17
1300 Mw	1.21	1.27

Table 9.2 implies that nuclear per-kwh capital charges are minimized at about 600-Mw unit size. In addition, there is only a 3%-4% deviation from this optimal size within the 400-850 Mw range. Per-kwh capital charges begin to increase sharply at larger unit sizes, however, since the declining benefits of capital economies of scale are increasingly outweighed by declining reliability. From Table 9.2, an 1150-Mw unit, which is the nuclear unit size most commonly being constructed and planned by U.S. utilities, would have per-kwh capital costs 13% and 17% higher than the minimum cost for PWR's and BWR's, respectively.

Of course, it is impossible to determine at present whether the inverse relationship between nuclear capacity factor and unit size will continue (see Section 9.6).

For coal-fired units, the variation in per-kw capital cost as a function of unit size is given by Equation 4.4, with k = 0.23. The ten-year levelized average capacity performance of coal-fired units, based on the CEP data base, is given by Equation 6.14 for supercritical units 600-Mw units or greater, and Equation 6.18 for subcritical units 400 Mw or greater.

(Equation 6.14)*

$$\text{Capacity Performance} = 84.21 - 2.394 \times (\text{Mw}/100)$$

(Equation 6.18)

$$\text{Capacity Performance} = 74.84 - 2.160 \times (\text{Mw}/100)$$

Equations 6.14 and 6.18, with Equation 9.4, give Table 9.3 of coal per-kwh capital charges as a function of unit size, using the capital cost and capacity factor of a 600-Mw coal unit as a reference point.

Table 9.3 shows the per-kwh capital charges for coal-fired units are lowest at 650 Mw for both subcritical and supercritical units. The per-kwh capital charge curve is fairly flat in the vicinity of these unit sizes, increasing by only about 2% within 225 Mw of the lowest-cost size for each boiler type.

Most coal units under construction are in the lowest-cost ranges shown in

*Equation 6.14 was chosen in preference to other trend equations for supercritical units. Equation 6.3 has less statistical significance; Equation 6.15 uses company variables which eliminate the effect of unit size. Use of either Equation 6.3 or 6.15 yields a considerably larger optimal unit size.

TABLE 9.3: RATIO OF COAL-FIRED PER-KWH CAPITAL CHARGES TO CAPITAL CHARGES OF A HYPOTHETICAL 600-MW COAL-FIRED UNIT†

Unit Size	Supercritical	Subcritical
300 Mw	1.06	1.06
400 Mw	1.03	1.03
500 Mw	1.01	1.01
600 Mw††	1.00	1.00
700 Mw	1.00	1.00
800 Mw	1.00	1.01
900 Mw	1.02	1.02
1000 Mw	1.03	1.03
1150 Mw	1.06	1.07
1300 Mw	1.11	1.11

†Table 9.3 assumes that the capacity factors of coal-fired units are given by their capacity performance.

††The unit sizes for which the per-kwh capital charge is minimized for coal units are 658 Mw for supercritical and 648 for subcritical.

Table 9.3. The major exception is the 1300-Mw supercritical units being planned and constructed by American Electric Power. However, AEP has thus far achieved successful reliability and therefore economical operation with its existing 1300-Mw units (see Section 9.6).

Figure 9.2 compares the per-kwh capital charges of PWR's, BWR's, and coal-fired units to 600-Mw units of each fuel and design type. Because the difference between subcritical and supercritical units is so slight, they have been combined.

Section 9.5: Other Factors Affecting Optimal Unit Size

The minimum capital charge unit sizes derived above are based on capacity performance as reflected in the CEP capacity performance trend equations and capital economies of scale estimated from the ERDA study of capital costs. Other factors relating to unit size affect per-kwh costs at the individual unit and system levels. These factors strengthen the conclusion that very large unit sizes are not economical, based on utility experience to date.

9.5a: Economies in Multi-Unit Stations

Some of the scale economies in construction costs which are foregone with smaller units may be recaptured when two or more identical units are built at the same site. Most of the design and engineering, licensing activity, site preparation and transportation facilities required for one unit can serve additional units at little additional cost, provided that construction of the second unit lags the first by only one or two years.* The net savings in the capital cost of the second

*Economies may also be realized through installation of common facilities larger than needed to serve the first unit alone. However, it is possible that the interest on the additional funds spent for such common facilities, applied during the period before completion of the second unit, may exceed the savings resulting from the lower direct cost (see Section 10.6).

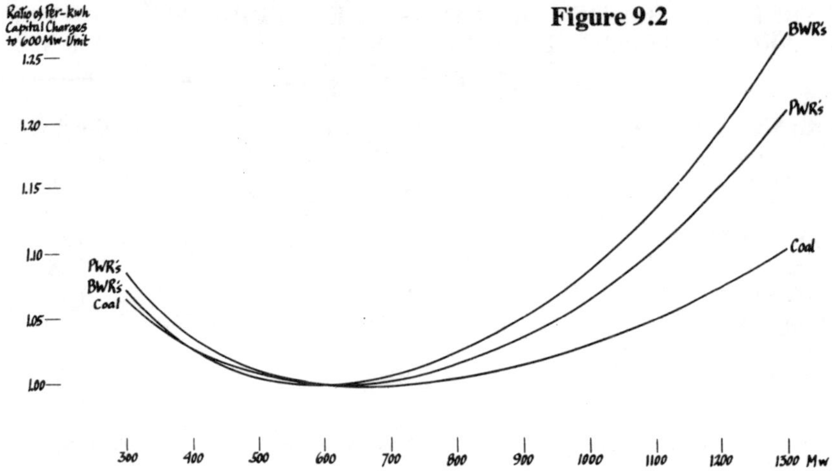

unit have been estimated in the ERDA study (WASH-1345) to be 10%-15%, or 5%-7% for an overall two-unit station.[3] This is consistent with estimates of such savings made by utilities in licensing applications. Such savings increase the economic advantage to installing several smaller units as an alternative to single, large units.*

Moreover, the operating reliability of identical second units installed at a site appears to be greater than that of comparable "non-duplicate" units. To date, large duplicate supercritical coal units have operated at an average capacity performance 5% higher than similar non-duplicate units. The corresponding figure for duplicate PWR's is 10%, although the data base for this result is very limited (see Equations 8.1 and 5.1). The potential higher capacity factor for duplicate units further increases the attractiveness of installing two or more smaller units at one site in lieu of one large plant.

9.5b: Thermodynamic Efficiency and Unit Size

A major factor in the trend to larger unit sizes for fossil plants was the increase in thermodynamic efficiency associated with larger sizes. The "heat rate" of new units (the number of Btu's of energy in fuel required to produce one kwh of net generation) decreased steadily in the past. This was partially due to technical improvements unrelated to unit size, but also due to efficiencies inherent in larger sizes of boilers and turbine-generators.

However, the scale economies in generating efficiency for fossil plants appear to end at about 500 Mw. The average unit size of the 18 most efficient U.S. coal-fired units in 1973 (measured by heat rate) was 509 Mw. Only three of the most efficient units were 700 Mw or larger, despite the existence in that year of roughly 25 700+Mw coal-fired units.[4]** The most efficient oil-and-

*Some utilities have proposed installation of two or more very large units (e.g., two 1150-Mw nuclear units). We have not studied whether substitution of four 575-Mw units would realize greater savings through facilities duplication than would be realized from two 1150-Mw units.

**The heat rates of the 18 units ranged from 8714 Btu/kwh (39% generation efficiency) to 9096 (38% efficiency).

gas-fired units tend to be smaller still. These figures strongly suggest that installation of very large coal-fired units will not increase, and may reduce, generation efficiency.

Extensive data is not readily available on generation efficiency for nuclear units, but the heat rates of current, commercial-sized nuclear units are generally assumed to fall within a relatively narrow range* and to be unrelated to unit size for units operating at high rates of capacity.

9.5c: Fuel Costs and Capacity Factor

A power plant's fuel costs per kwh are not independent of its capacity factor. For fossil units, operation at low load reduces generating efficiency and therefore increases per-kwh fuel costs. This is primarily because heat losses are relatively constant at any load and are therefore more significant at low load levels.**

In nuclear plants, low capacity factors may contribute to higher fuel costs by reducing thermal efficiency and by causing nuclear fuel to be burned to less than its design level at the time that it must be removed from the plant. Utilities purchase nuclear fuel at specified "enrichment" levels (concentration of fissionable uranium) according to an anticipated quantity of electricity to be generated by the fuel before the unit is shut down for refueling. In theory, the additional cost of the high level of enrichment usually specified by utilities is more than recovered through additional operation of the nuclear unit in lieu of other units with higher incremental costs.***

In practice, however, most nuclear units have operated at capacity factors lower than anticipated and in many cases have discharged a significant inventory of unused fuel during refueling outages. Although the unused fuel may ultimately be converted to re-usable form in reprocessing plants (and in fact utilities generally take a credit for the future value of such fuel in their nuclear fuel accounting), the discharge of unused nuclear fuel adds to nuclear fuel costs to some extent.

Accordingly, to the extent that larger nuclear plants are more prone to outages which reduce capacity factor, they are more likely to sustain higher fuel costs through reduced utilitization of purchased nuclear fuel.****

*Roughly 10,200-11,000 Btu/kwh heat rate, corresponding to an efficiency of 31%-33%. Some nuclear units operate at average heat rates higher than 11,000 Btu/kwh.

**Frequent deratings of fossil units would be expected to reduce their thermal efficiency, perhaps accounting in part for the scarcity of very large units—with their generally lower capacity factors—in the ranks of most efficient units.

***The higher the level of enrichment, the higher the cost of a batch of nuclear fuel, due to the higher quantity of enrichment work required and the larger quantity or uranium concentrate ("yellow-cake") necessary to produce a given quantity of uranium enriched to the higher level.

****Nuclear fuel "management," as the design and operation of reactor cores is known, is a complex subject which lies beyond the scope of this study. Several points are worth noting, however. In order to ensure availability of maintenance personnel to expedite refueling and concurrent repairs and to have the unit available for service during peak-demand seasons utilities have a strong incentive to refuel nuclear units at pre-arranged intervals. While some flexibility is often built into utilities' refueling schedules, such flexibility may be limited by these constraints. Thus, a hypothetical nuclear fuel core designed to achieve sufficient burnup to generate 5 million

9.5d: O& M Costs and Unit Size

Operating and maintenance costs account for about 10% of the annual generating costs of nuclear and coal-fired units. O&M costs are usually estimated on an annual basis geared to the unit's size and type.

O&M costs for coal-fired units with scrubbers are anticipated to be higher than for conventional coal units. This is due to higher annual operating and maintenance requirements, and to variable (generation-dependent) costs for the chemical reagent (such as limestone) required and the disposal of sludge or other byproducts.

Utilities often project lower per-kwh O&M costs for large units. This would appear to be a reasonable assumption for operating costs, since fewer personnel per kw should be required to operate large units. Maintenance costs, however, could be higher per kwh generated for large units, with their lower average capacity factor and potentially greater maintenance and repair requirements associated with units undergoing more frequent outages. Based on the inverse relationship of capacity factor to unit size for nuclear units, and coal units to a lesser extent, significant economies in per-kwh O&M costs are not likely to be achieved by installing large unit sizes.

9.5e: Unit Size and Reserve Margins

Utilities plan and construct new generating facilities so that their total installed capacity equals anticipated "peak load" (customer demand plus transmission line losses) plus sufficient reserve capacity to meet demand regardless of generating equipment outages. Consequently, "reserve margin

---

Mwhrs in the 12 months between refuelings (assuming a 90% capacity factor between refuelings) will generate 3.3 million Mwhrs during the same 12 months if it is limited by outages to an average 60% capacity factor. Extending the cycle by three months will bring the per-cycle generation to 4.2 million Mwhrs at a 60% capacity factor, but at this point the unit may have to be refueled because of scheduling constraints. The result is that a portion of the fuel is discharged at burnup level less than the design level.

Of course, utilities could design and purchase low-enriched nuclear fuel in anticipation of low capacity factors. This is impractical, however, due to the difficulty of predicting capacity factors during a given operating cycle. A unit with fuel designed for a low capacity factor must either be operated at a low rate of capacity or, if its availability turns out to be high, must be refueled ahead of schedule. This may create problems at the system level.

Low overall nuclear plant reliability, and instances of defective fuel in particular, have caused the average burnup of nuclear fuel to be considerably below design levels.[5] Nevertheless, each of the half-dozen of so utility applications to build nuclear units which CEP has examined in depth has projected nuclear fuel costs on the basis of high design burnups predicted by fuel vendors. The expectation by utilities that these burnup levels will be achieved in new facilities requires that new nuclear units operate at high (75%-80%) capacity factors.

criterion'' (the amount of reserve capacity deemed necessary for reliable utility system operation) is directly affected by the reliability of the individual units. To the extent that smaller unit sizes are more reliable, they will reduce a system's reserve margin criterion.* This would reduce the overall need for new generating facilities and, commensurately, total system generating costs.

In addition, a utility system or power pool with large generating units may tend to require more reserve capacity than a system with smaller units, independent of the poorer average reliability of large units, since the loss of a single large unit will reduce the system's reserve margin more than the loss of one small unit.**

9.5f: Unit Size and Siting Economies

The choice of smaller unit sizes may further reduce overall utility costs by creating more choices in site selection. This consideration could be especially applicable to coal-fired plants since they must be projected to meet ambient air quality requirements as a prerequisite for a construction permit. The potential advantages of flexibility in siting include reduction in transmission line costs and energy losses, access to larger markets for industrial or commercial sale of low-pressure steam, and reduction in utility management time expended in site selection.

9.5g: Time Required for Construction

The capital cost scaling factors for nuclear and coal plants derived from the ERDA study are based primarily upon vendor and utility estimates of scale economies for categories of plant equipment such as the nuclear steam supply system, fossil boiler, or turbine-generator set. The overall scaling factors estimated by ERDA (see Sections 9.3 and 9.4) thus do not appear to allow for potential variations in the duration of construction as a function of plant size. To the extent that large plants may require a longer construction period than small ones (thereby accumulating greater interest costs during construction), their capital economies of scale may be overstated in the ERDA study.

Section 9.6: Future Nuclear and Coal Plant Sizes

Section 9.4 provided estimates of unit sizes for which nuclear and coal plants' capital charges per kwh are minimized, based on the CEP capacity performance trend equations and the ERDA capital cost economies of scale (see Table 9.4). Section 9.5 cited factors which would tend to cause the system-optimal unit size to be less than the plant capital charge-minimizing unit size.

*Moreover, percent of total unit downtime attributable to forced outages appears to be an increasing function of unit size.

**This consideration, however, generally does not apply in cases where the largest unit in the system or pool accounts for 7% or less of the system-wide capacity. In such cases, system capacity is distributed sufficiently well among different units that loss of a single or several units due to forced outages will not impair system reliability much more than the loss of a greater number of smaller units.[6]

TABLE 9.4: OPTIMAL UNIT SIZE RANGE FOR NUCLEAR AND COAL†

Unit Type	Unit Size for Minimum Capital Cost per-kwh	Unit Size Range for Minimum System Cost
PWR	622 Mw	350-700 Mw
BWR	585 Mw	350-650 Mw
Supercritical coal	658 Mw	350-800 Mw
Subcritical coal	648 Mw	350-800 Mw

†Column 3 in Table 9.4 was derived by estimating the Mw range for which each unit type's per-kwh capital charges would be no more than 2% greater than that for the cost-minimizing unit size in column 2 (see Tables 9.2 and 9.3). This range was then reduced by 100 Mw, to account for the estimated effect of other factors affecting optimal unit size (see Section 9.5).

The ranges in column 3 are wider for coal than for nuclear units because of the flatter slope of the capacity performance trend equations for coal.

Accordingly, the optimal size of new units would tend to be even less than the size at which capital charges are minimized, as shown in Table 9.4.

From Table 9.4, the optimal size range for new units is about 350-700 Mw for nuclear and 350-800 Mw for coal. These figures assume that the relationships between capacity factor and unit size in the CEP trend equations are followed by new units.

Significantly, unit sizes of most coal-fired units being constructed by U.S. utilities appear to be within the optimal range estimated, while new nuclear units will be much larger than their optimal size range (see Table 9.5). Of 364 new fossil units, only five will be 900 Mw or larger. Two will be 1300 Mw units of American Electric Power, which has already successfully operated three such units. The average size of the new fossil units is 457 Mw. Moreover, the 350-800 Mw range, which is approximately the optimal size range for coal, accounts for 61% of new units and 73% of new capacity.

Conversely, only three of 145 new nuclear units are within 350-700 Mw, the range indentified as economically optimal for nuclear plants. Two others will be 700-799 Mw and the remainder are 800+ Mw. The average capacity of all 145 new nuclear units is 1064 Mw.* At this capacity, the per-kwh capital charges for a PWR are 9% greater than for a 600-Mw unit. BWR's are 12% greater.

These figures pertain only to per-kwh capital charges and do not reflect the other potential dis-economies of large unit sizes (see Section 9.5). Moreover, 57% of new nuclear units will be 1100+ Mw and so will further deviate from optimum nuclear unit size. By comparison, the average nuclear unit in the CEP data base is only 744 Mw and the average nuclear unit-year is 681 Mw.

Most utilities defend their selection of large unit sizes for new nuclear plants on the basis of anticipated capacity factor improvements as such units mature

*The figures on nuclear unit size understate the average capacity of units yet to be installed, since figures include units completed in 1975. Six of the ten nuclear plants declared commercial in 1975 are less than 900 Mw capacity. The figures exclude the 330-Mw Fort Saint Vrain high-temperature gas-cooled reactor and the 300-Mw Clinch River liquid metal-fast breeder reactor. Neither is a light-water reactor, the reactor type studied here.

TABLE 9.5: FOSSIL AND NUCLEAR STEAM UNIT SIZE RANGES[7]

PROPOSED GENERATING CAPACITY ADDITIONS FOR THE PERIOD 1975 THROUGH 1984

Unit Size	Fossil Steam				Nuclear Steam			
	Capacity		Units		Capacity		Units	
Range (MW)	MW	%	No.	%	MW	%	No.	%
Below 100	1,247	0.75	29	7.97	0	0.00	0	0.00
100-199	4,792	2.88	33	9.07	0	0.00	0	0.00
200-299	8,723	5.25	37	10.16	0	0.00	0	0.00
300-399	11,191	6.73	33	9.07	630	0.41	2	1.36
400-499	18,869	11.35	44	12.09	482	0.31	1	0.68
500-599	41,397	24.90	79	21.70	0	0.00	0	0.00
600-699	30,258	18.20	47	12.91	1,268	0.82	2	1.36
700-799	25,755	15.50	35	9.62	1,540	0.99	2	1.36
800-899	18,319	11.02	22	6.04	15,903	10.26	19	12.93
900-999	910	0.55	1	0.27	21,109	13.62	23	15.65
1000-1099	1,060	0.64	1	0.27	14,919	9.63	14	9.52
1100-1199	1,100	0.66	1	0.27	54,725	35.32	48	32.65
1200-1299	0	0.00	0	0.00	41,762	26.95	34	23.13
1300-1399	2,600	1.56	2	0.55	2,600	1.68	2	1.36
Total	166,221	100.00	364	100.00	154,938	100.00	147	100.00

and as architect-engineers (and utilities themselves) develop design and operating experience with large units as a whole.

That very large coal-fired generating units can be built and operated reliably appears to be borne out by the experience of American Electric Power Co.'s three new 1300-Mw supercritical coal-fired units. The performance of these units has matched the generally high reliability of AEP's supercritical 615-Mw and 816-Mw units, although the three unit-year total is not an adequate basis for drawing firm conclusions (see Section 7.3).

AEP attributes the high reliability of the 1300-Mw units to conservative design and a series concept of standardized design application to larger units. It is beyond the scope of this study to say whether the many new very large nuclear units, and fewer very large coal units, follow a design philosophy comparable to AEP's. However, it does appear that nuclear units have been sized at increasing capacities without the full benefit of successful experience at intermediate sizes such as enjoyed by AEP. Moreover, it is possible that nuclear plants are inherently more prone to reliability problems at larger unit sizes than coal-fired units due to the greater tendency to shut nuclear plants because of component failure.

Notwithstanding AEP's successful experience at 1300 Mw, utilities which take credit in their cost estimates for economies of scale in construction costs of large units should seriously consider whether they may realistically anticipate high capacity factor returns from large units.

REFERENCES

1. The major reports published under the study are the AEC's report WASH-1345, "Power Plant Capital Costs: Current Trends & Sensitivity to Economic Parameters," published in October 1974 and still the most recent federal report on nuclear and coal capital costs; and ERDA-108, "CONCEPT—A Computer Code for Conceptual Cost Estimates of Steam-Electric Power Plants—Phase IV, User's Manual," June 1975.

Our estimates of k are based on WASH-1345, Figure 10, p. 44, and fit the curves in Figure 10 with an error of less than 1%. Figure 10 covers a size range of 600-1300 Mw; we have extrapolated the cost-size relationship to 400 Mw.

2. ERDA-108, *op. cit.*, p. 14.

3. WASH-1345, *op. cit.*, p. 74.

4. Federal Power Commission, "Steam-Electric Plant Construction Cost and Annual Production Expenses," 1973, p. xxxv.

5. See, for example, Jim Harding, "Nuclear Fuel Burnup and Fuel Cycle Costs," California Energy Resources Conservation & Development Commission, June 1976.

6. See, for example, L.K. Kirchmayer, A.G. Mellor, J.F. O'Mara, and J.R. Steveson, "An Investigation of the Economic Size of Steam-Electric Generating Units," *AIEE Transactions*, 74(3), August 1955, pp. 600-609.

7. "Fossil & Nuclear Steam Unit Size Range—Proposed Generating Capacity Additions for the Period 1975 through 1984," as reported by the nine Regional Electric Reliability Councils to the FPC on April 1, 1975, printed in *FPC News*, December 9, 1975. The size distribution of fossil units is assumed to represent that of coal units as well. This appears to be a reasonable assumption since 81% of the new fossil units reported by the FPC in the same report will burn coal or lignite or will have the capability of doing so.

10: Economic Evaluation of Nuclear and Coal-Fired Units

Section 10.1: Introduction and Key Findings

The relative generating costs of new nuclear and coal-fired units are currently the subject of much study and debate. The costs have been analyzed by electric utilities, generating equipment vendors, architect-engineers, and more recently by critics of nuclear power who have argued that nuclear plants are not economically superior to coal-fired generation. Some authorities insist that the choice in an economy with diminishing oil and gas resources is not "nuclear vs. coal" but "how much of each." Still, comparative analysis of nuclear and coal costs in separate states and utility service territories is significantly affecting U.S. electric supply planning.[1]

CEP findings on the generating costs of nuclear and coal-fired units are based on various assumptions of unit size (600 Mw vs. 1150 Mw) and design: pressurized vs. boiling water reactor, supercritical vs. subcritical boiler, scrubber vs. low-sulfur coal. Standard assumptions are used for variables not directly related to unit design or size—fuel costs, operating and maintenance costs, and base capital costs—so that the effect of size variations on capacity factor and capital cost may be demonstrated. Capacity factor and capital cost are assumed to vary according to the capacity factor trend equations (see Chapters 3 and 6) and the ERDA capital economies of scale (see Chapter 9).

The following are our major economic findings:

1. New 1150-Mw PWR's and BWR's will be 16% and 31% more expensive, respectively, than 600-Mw supercritical coal units using Eastern low-sulfur coal, without scrubbers. If scrubbers rather than low-sulfur coal are used 1150-BWR's will be only 12% more expensive. PWR's will then equal the Eastern coal unit cost. *This is based on projections from the CEP capacity factor trend equations and standard cost assumptions for other parameters.*

This contradicts standard utility assumptions of 75% nuclear and coal capacity factors (71% for coal with scrubbers), under which new 1150-Mw nuclear units would have the lowest generating costs for an Eastern U.S. site. Coal

TABLE 10.1: 1984 NUCLEAR AND COAL GENERATING COSTS†

Eastern Coal	Generating Cost (Mills/Kwh)	Ratio to 1150-MwPWR
Low-Sulfur		
Mine-Mouth	41.3	0.89
Transported	44.8	0.97
High-Sulfur		
Mine-Mouth	48.6	1.05
Transported	51.9	1.12
Western Coal		
Mine-Mouth	41.8	0.85
Transported	45.5	0.93

†Costs assume capacity factors from CEP trend equations, with 7.5 percentage points added for nuclear.

Eastern coal case requires low-sulfur coal or scrubbers, but not both. Costs are as in Table 10.4, except that variable scrubber O&M cost is reduced 20% for mine-mouth plant to reflect cheaper scrubber feed and disposal costs. PWR cost for Eastern case is 46.2 mills; BWR cost is 50.9 mills.

Western coal is low-sulfur but still requires scrubbers. Costs are as in Table 10.5. PWR cost is 49.0 mills; BWR cost is 54.0 mills.

units of 600 Mw would be 23%-37% more expensive.*

2. When no improvement in coal capacity factors is assumed, but a nuclear improvement 7.5 percentage points over CEP capacity factor trend equations is granted on the assumption of improvements in plant design and operation, the generating costs of 1150-Mw nuclear and 600-Mw supercritical Eastern coal units fall in a narrow range. Low-sulfur Eastern coal without scrubbers is cheapest. PWR's add 3%; BWR's add 14%; high-sulfur coal with scrubbers adds 17%.

3. For a Western site, 1150-Mw PWR's are 8% and BWR's are 19% more expensive than 600-Mw supercritical coal units. This assumes: coal units with scrubbers; use of low-sulfur coal; all units with dry cooling towers; and a 7.5% capacity factor improvement for nuclear units.

4. Cost comparisons assume coal must be transported from mine to plant. Relative coal-to-nuclear cost is reduced by about 10% for mine-mouth plants.

5. "Early installation" of very large nuclear plants—construction prior to their need for adequate system reliability—will not create overall cost savings because of low capacity factors (under 55%).

Table 10.1 summarizes relative costs of 600-Mw supercritical coal units and 1150-Mw PWR's. These are the most likely to be used designs and sizes for new coal and nuclear plants.

Section 10.2: Cost Assumptions

This section sets forth assumptions used in calculating total nuclear and coal generating costs. Since many of the cost parameters are currently under debate,

*The 23% figure assumes use of Eastern low-sulfur coal without scrubbers. The 37% would be with scrubbers.

a middle ground has been chosen. Assumptions pertain to single-unit stations in the East and use 1984 as a reference point.* Costs for Western plants and mine-mouth sites are discussed elsewhere (see Section 10.5).

A. Capital Costs

1. The cost ratio of scrubber-equipped coal units to nuclear is equal to that in the AEC's WASH-1345 report (78% at 600 Mw, 82% at 1150 Mw).[2]
2. Coal units with scrubbers have capital costs 15% greater than those without.
3. The level of costs is determined by taking 1981 costs in WASH-1345, escalated at 6% per year for three years and increased 50% to compensate for AEC's historic underestimate of capital costs. The resulting costs for 1150-Mw units—$1020/Kw for nuclear, $839/Kw for coal with scrubbers, $729/Kw for coal without scrubbers—are consistent with most mid-1980's estimates.
4. PWR/BWR costs are equal, as are supercritical/subcritical costs. Similarly, there is no differentiation between unit designs for estimating fuel and O&M costs.
5. Annual fixed charge rate is 16%.

B. Operating and Maintenance Costs

1. O&M costs for 1150-Mw units are based on estimates for 1984 by Northeast Utilities.[3] They are: nuclear, $20.63/Kw-year; coal without scrubbers, $16.51/Kw-year; coal with scrubbers, $20.44/Kw-year, plus a variable component of 1.81 mills/kwh.**
2. For 600-Mw units, annual O&M costs are 10% greater than for 1150 Mw. The variable cost component for scrubbers is unchanged.

C. Fuel Costs

1. Eastern high-sulfur coal, 12,000 Btu/lb, costs $23.50 per ton at the mine in 1975, plus $7/ton for transportation. Escalating at 5% per year, the 1984 cost is $1.97 per million Btu.
2. Eastern low-sulfur coal, with equal Btu content, has a 20% greater mine cost than high-sulfur coal and equal transportation cost. The 1984 cost is $2.28 million Btu.***

*For simplicity, we have projected costs for a single year rather than performing a 10-year levelized analysis (except insofar as capacity factor projections are based on 10-year levelized averages). This assumption tends to favor coal over nuclear, due to coal's greater variable cost component. However, the effect is quite small, especially insofar as "steady-state" nuclear fuel costs are applied, whereas nuclear fuel costs are actually high during the first several years of operation; and insofar as the increase in nuclear fuel costs may exceed that for coal after 1984.

**These costs are consistent with or more favorable to nuclear than many other studies.[4]

***The high-sulfur 1975 mining cost, 1975 transportation cost, and heat content are from T.H. Lee,[4] but the 5% escalation rate is less than his 8%-9% estimate.

The cost and availability of Eastern low-sulfur coal are based upon CEP's study of BuMines reports on Eastern deep-mine coal reserves by sulfur content, current and projected cost differen-

3. The heat rates of coal-fired units are 9600 Btu/Kwh with scrubbers, 9100 Btu/Kwh without. This corresponds to 36% and 38% thermal efficiency, respectively.
4. Nuclear fuel costs are 8 mills/Kwh in 1984. This is consistent with 1984 uranium yellowcake costs of $50/lb, enrichment costs of $120-$140 per separative work unit, and spent fuel reprocessing at $225 per kilogram to recover salable uranium and plutonium. Nuclear fuel costs are computed using the vendors' design burnup projections for the more efficient steady-state case rather than initial core operation, and are assumed to be independent of unit size.[6]

Section 10.3: Future Generating Costs Under Utility and CEP Capacity Factors

Besides assumptions for capital, fuel, and O&M costs, computations for nuclear and coal generating costs rely on two sets of capacity factor assumptions:

a) A "utility base case" of 75% capacity factor for nuclear and coal-fired units without scrubbers, 71% for coal units with scrubbers. This case assumes that new nuclear and coal units will achieve a high degree of operating reliability; that capacity factor is not affected by unit size or coal sulfur content; and that scrubbers will decrease coal-unit reliability by five percent, so that units with scrubbers have a capacity factor of 71%.

b) A "CEP data case" in which capacity factors are determined by applying Trend Equations 3.2, 3.3, 6.12, and 6.16, and in which scrubbers again reduce coal unit capacity factors by 5%. This case assumes that capacity performance of future nuclear and coal plants will follow patterns now established and that coal-fired units will be fully base-loaded so their capacity factors and capacity performance will be equal. Overall costs of various nuclear and coal alternatives for the two cases are set forth in Table 10.2.

Generating costs for the different cases in Table 10.2 are summarized in Table 10.3.

Under the utility base case new 1150-Mw coal-fired units meeting federal EPA emission standards would be 14%-27% more expensive to own and operate than 1150-Mw nuclear units—the typical size of new nuclear units proposed by utilities. Coal-fired units of 600 Mw would be 23%-37% more expensive than 1150-Mw nuclear units. The lower figures in the percentage ranges pertain to coal-fired units using Eastern low-sulfur coal.*

Using the CEP data case (Part B of Table 10.3), it is nuclear units that are more expensive to own and operate than the best coal alternative—600-Mw

tials between high- and low-sulfur coal, and costs of coal cleaning processes. The reserve base of low-sulfur coal in West Virginia and eastern Kentucky in coalbeds 28+ inches thick, maximum depth 1000 feet, can support the lifetime operations of 120,000 Mw of new coal-fired units, without scrubbers. This is at 50% mining recovery rates, without the use of coal cleaning processes.[5]

*Utility cost estimates of Eastern coal units assume scrubber use, without analyzing availability of low-sulfur Eastern coal. Western low-sulfur coal is not generally cheaper than Eastern coal with scrubbers, due to high West-to-East shipping cost. Utility coal availability and nuclear capacity factor assumptions suggest new coal unit generating costs will be 26%-37% more expensive than 1150-Mw nuclear units, depending on coal unit size.

TABLE 10.2: 1984 GENERATING COSTS OF EASTERN NUCLEAR AND COAL-FIRED UNITS UNDER DIFFERENT CAPACITY FACTOR ASSUMPTIONS†

A1. UTILITY BASE CASE, 600 Mw

	Nuclear	Low-Sulfur Coal†† (no scrubbers)	High-Sulfur Coal†† (scrubbers)
Capital Cost: $/Kw	1256	847	974
Annual Cap Chrg: $/Kw-yr	201	136	156
O&M: $/Kw-yr	23	18	34
Capacity Factor	0.75	0.75	0.71
Capital + O&M: mills/Kwh	34.1	23.4	30.5
Fuel: mills/Kwh	8.0	20.7	18.9
TOTAL: mills/Kwh	42.1	44.1	49.4

†Comparisons of generating costs in mills/Kwh for generating alternatives with different capacity factors may result in some error since alternatives are being compared on the basis of different quantities of Kwh's generated. Standard practice is to add the cost of replacement power to the low-capacity factor alternative and to compare the alternatives on the basis of equal generation. Furthermore, replacement power costs may be different for scheduled vs. forced outages. However, the error in the table is small.

††High-sulfur coal is assumed to be 3%-sulfur coal; low-sulfur coal is less than 1%-sulfur (see Equation 6.12).

A2. UTILTY BASE CASE, 1150 Mw

	Nuclear	Low-Sulfur Coal†† (no scrubbers)	High-Sulfur Coal†† (scrubbers)
Capital Cost:$/Kw	1020	729	839
Annual Cap Chrg: $/Kw-yr	163	117	134
O&M: $/Kw-yr	21	16	32
Capacity Factor	0.75	0.75	0.71
Capital + O&M: mills/Kwh	28.0	20.2	26.7
Fuel: mills/Kwh	8.0	20.7	18.9
TOTAL: mills/Kwh	36.0	40.9	45.6

B1. CEP DATA CASE, 600 Mw

	Nuclear		Low-Sulfur Coal (no scrubbers)		High-Sulfur Coal (scrubbers)	
	BWR	PWR	Sub-Critical	Super-Critical	Sub-Critical	Super-Critical
Capital Cost: $/Kw	1256		847		974	
Annual Cap Chrg: $/Kw-yr	201		136		156	
O&M: $/Kw-yr	23		18		32	33
Capacity Factor	59.6%	66.2%	65.4%	72.8%	62.2%	64.6%
Capital + O&M: mills/Kwh	42.9	38.0	26.9	24.1	34.5	33.4
Fuels: Mills/Kwh	.8.0		20.7		18.9	
TOTAL: mills/Kwh	50.9	46.6	47.6	44.8	53.4	52.3

B2. CEP DATA CASE, 1150 Mw

	Nuclear		Low-Sulfur Coal (no scrubbers)		High-Sulfur Coal (scrubbers)	
	BWR	PWR	Sub-Critical	Super-Critical	Sub-Critical	Super-Critical
Capital Cost: $/Kw	1020		729		839	
Annual Cap Chrg: $/Kw-yr	163		117		134	
O&M: $/Kw-yr	21		16		28	29
Capacity Factor	41.5%	47.6%	47.4%	59.7%	45.0%	52.1%
Capital + O&M: mills/Kwh	50.6	44.1	32.0	25.4	41.1	35.7
Fuel: mills/Kwh	8.0		20.7		18.9	
TOTAL: mills/Kwh	58.6	52.1	52.7	46.1	60.0	54.6

TABLE 10.3: 1984 GENERATING COSTS OF EASTERN NUCLEAR AND COAL-FIRED UNITS, IN MILLS/KWH (AND RATIO TO 1150-MW NUCLEAR COSTS UNDER UTILITY BASE CASE)

	Nuclear		Low-Sulfur Coal†† (no scrubbers)		High-Sulfur Coal†† (scrubbers)	
	BWR	PWR	Sub-Critical	Super-Critical	Sub-Critical	Super-Critical
A. Utility Base Case						
1. 600 Mw	42.1 (1.17)		44.1 (1.23)		49.4 (1.37)	
2. 1150 Mw	36.0 (1.00)		40.9 (1.14)		45.6 (1.27)	
B. CEP Data Case						
1. 600 Mw	50.9 (1.41)	46.6 (1.29)	47.6 (1.32)	44.8 (1.24)	53.4 (1.48)	52.3 (1.45)
2. 1150 Mw	58.6 (1.63)	52.1 (1.45)	52.7 (1.46)	46.1 (1.28)	60.0 (1.67)	54.6 (1.52)

supercritical units burning Eastern low-sulfur coal. PWR's of 600 Mw are 4% more expensive, BWR's 14% more. At the planned 1150-Mw nuclear unit size, PWR's would be 16% and BWR's 31% more expensive than the best coal alternative.

If 1150-Mw nuclear units are assumed, along with 600-Mw coal units with scrubbers, PWR's cost will equal supercritical coal units. BWR's will be 12% more expensive than either, based on CEP capacity factor equations. At the 600-Mw level, subcritical coal units will be slightly more expensive than supercritical: 2% more if the coal units are scrubber-equipped, 6% more if they are not.

PWR's of 1150 Mw and supercriticals of 600 Mw are the most likely design and size choices for new nuclear and coal units. The projected ratio of coal-to-nuclear generating cost shifts from 1.23 under the utility case to 0.86 under the CEP case, without scrubbers, and from 1.37 to 1.00 with them. The changes in ratios are greater for BWR's. The 37 percentage point difference in relative cost estimates demonstrates the impact of capacity factor upon power plant generating costs. It also dramatizes the difference between the optimistic utility view and that of observers who are skeptical of those new unit operating reliability predictions that greatly exceed historical patterns.

The great variability of generating costs with changes in capacity factor is indicated in Figure 10.1.

Section 10.4: Future Generating Costs Assuming Nuclear Capacity Factor Improvement

It is realistic to expect improvements in design and operating practices to increase nuclear and coal capacity factors. Problem areas such as radioactive

Figure 10.1

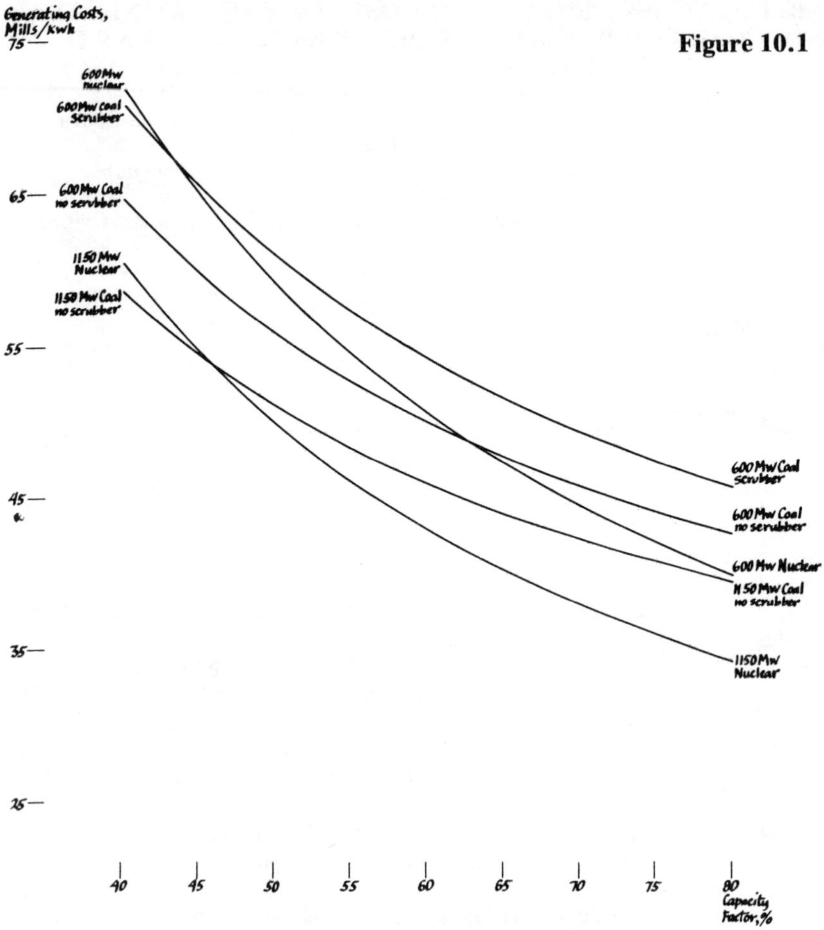

Note: Generating costs of 1150-Mw coal plant with scrubber not shown, since such costs are identical (within 0.1 mills/kwh) to costs of 600-Mw coal plant without scrubber.

exposure to maintenance personnel in nuclear units and changes in fuel quality at coal units, for example, may be better controlled by new design practices. Design duplication and standardization—particularly for nuclear plants—should also improve operating reliability.

It is also clear that technical complexity of nuclear and coal units and strict regulation of plant operations, especially nuclear, will limit reliability improvements (see Sections 5.5 and 8.4). Therefore new units, especially very large ones, will not likely operate at the 75% utility base case capacity factor prediction.

The overall cost analysis assumes very large (1150-Mw) nuclear units will operate at capacity factors 7½ percentage points higher than predicted from CEP trend equations. However, capacity factors of 600-Mw supercritical coal

TABLE 10.4: 1984 GENERATING COSTS OF EASTERN NUCLEAR AND COAL-FIRED UNITS, ASSUMING 7½% INCREASE IN NUCLEAR CAPACITY FACTORS

	1150-Mw BWR	1150-Mw PWR	600-Mw Super-critical, Low-sulfur Coal	600-Mw Super-critical, Scrubbers
Capital Cost: $/Kw	1020	1020	847	974
Annual Cap Chrg: $/Kw-yr	163	163	136	156
O&M: $/Kw-yr	21	21	18	33
Capacity Factor	49%	55%	72.8%	64.6%
Capital + O&M: mills/Kwh	42.9	38.2	24.1	33.4
Fuel: mills/Kwh	8.0	8.0	20.7	18.9
TOTAL: mills/Kwh	50.9	46.2	44.8	52.3
Ratio to Low-Sulfur Coal Case	1.14	1.03	1.00	1.17

units are assumed to be as predicted from the equations.* Table 10.4 calculates generating costs of 1150-Mw nuclear and 600-Mw supercritical coal units in the East.

1150-MwPWR's are 3% and BWR's are 14% more expensive than 600-Mw supercritical coal units without scrubbers. If low-sulfur coal is not available, however, then the alternative coal unit with scrubbers would cost 13% more than PWR's and 3% more than BWR's. Overall, coal-fired units are competitive with nuclear plants in the East, even assuming improved nuclear capacity factors and standard capital, fuel and O&M costs of Table 10.4.

Section 10.5: Nuclear and Coal Costs for Different Geographical Cases

For a given utility, the standardized parameters in this analysis should be examined on a site-specific basis. Capital costs are sensitive to geographical

*The assumed improvement for nuclear but not for coal reflects (i) a correction for applying the CEP capacity factor trend equations to the projected 1150-Mw nuclear units, since the negative relationship between capacity factor and unit size for the 450-Mw to 1098-Mw nuclear units in the data base may be reduced outside of that range; (ii) the fact that application of Equation 6.13 rather than Equation 6.12 would reduce the projected capacity performance of supercritical units by several percentage points; and (iii) an effort to be "conservative" (favorable to nuclear) in the nuclear and coal cost analysis.

TABLE 10.5: 1984 GENERATING COSTS FOR NUCLEAR AND COAL-FIRED UNITS IN WESTERN U.S.

	1150-Mw BWR	1150-Mw PWR	600-Mw Supercritical, Low-Sulfur Coal+ Scrubbers
Capital Cost: $/Kw	1091	1091	1074
Annual Cap Chrg: $/Kw-yr	175	175	172
O&M: $/Kw-yr	21	21	31
Capacity Factor	49%	55%	69%
Capital + O&M: mills/Kwh	45.7	40.7	33.6
Fuel: mills/Kwh	8.3	8.3	11.9
TOTAL: mills/Kwh	54.0	49.0	45.5

variations in labor costs and productivity. Nuclear capital costs may also vary with site-related factors such as seismic design, aircraft hardening, and flood protection. Factors such as the utility's previous operating record with nuclear and coal units should be considered in projecting future capacity factors. These variations are beyond the scope of this study.

Table 10.5 compares nuclear and coal costs for a hypothetical Western U.S. site, using average regional assumptions about cooling water availability, environmental protection requirements, and coal fuel costs.

Regional variables that affect cost assumptions in Table 10.5 are:

1. Dry cooling towers add 7% to nuclear and 5% to coal capital cost.
2. Larger boiler to accommodate sub-bituminous coal adds an additional 5% to coal capital cost.
3. Coal unit uses scrubbers in addition to low-sulfur coal to comply with "Significant Deterioration" provision of the Clear Air Act.
4. Coal variable O&M cost (for scrubber feed and disposal) is 20% less than in Eastern case.
5. Coal fuel cost is $8/ton at mine plus $5/ton average transportation in 1975, escalated to 1984 at 5%, with heating value of 8500 Btu/lb. Transportation may be less in Rocky Mountain states, slightly higher on Pacific Coast.
6. Coal heat rate is 10,000 Btu/lb, reflecting lower efficiency with dry cooling tower. Similarly, nuclear fuel cost is increased 4% to 8.3 mills.
7. Capacity factor of the coal-fired unit is greater than for the scrubber case in Table 10.3, due to the lower sulfur content of Western coal.

Western generating costs put PWR's 8% and BWR's 19% higher than those for supercritical coal units using scrubbers and low-sulfur coal.

The costs in Tables 10.2 through 10.5 assume that coal must be transported from mine to plant. That comes to a 1984 average cost of about $11/ton (4.2 mills/Kwh) in the East and $8/ton (4.6 mills/Kwh) in the West. Table 10.1 gives costs of the most likely new plants (1150-Mw nuclear, 600-Mw supercritical coal) for mine-mouth coal plants that have minimal transportation cost

($1/ton in 1975). Relative coal-to-nuclear costs are reduced by about 10% for this case.

Section 10.6: Capacity Factors and "Early Installation" of Nuclear Units

Several East and West Coast utilities which are heavily dependent upon oil as a fuel for electric generation, have asserted that it is economical to install nuclear plants in advance of the date they will be needed for system reliability. They claim that fuel cost savings from installing nuclear plants to retire oil-fired generation outweigh additional fixed charges associated with carrying excess generating capacity.[7]

The merits of early installation* are dependent upon a number of factors: (i) nuclear fuel and O&M costs; (ii) fuel and O&M costs of oil-fired units reduced in load by operation of the nuclear unit; (iii) escalation rate for nuclear capital costs; (iv) time frame for the analysis; (v) discount rate for converting future costs to a common reference basis; (vi) capacity factor of the nuclear unit.** Most critical is the nuclear unit capacity factor, since fuel cost savings from early installation are directly proportional to kilowatt-hours generated.

The relatively low anticipated capacity factors for new nuclear units, even using adjusted capacity factors in Section 10.4, rather than capacity factors computed from CEP trend equations, reduce cost savings to the point where early installation is uneconomical, or marginal at best. This conclusion is based upon CEP analyses of cases in New York State, Connecticut, and New Jersey.[8]

In each case, a detailed year-by-year comparison was made of the utility's total fixed charges and fuel and operating costs under two alternative scenarios: (i) completion of a nuclear unit at an early date, in advance of its need for system reliability, and (ii) deferral of commencement and completion of the unit*** until the projected date that it will be needed to ensure adequate capacity reserve.

The following is a summary of analysis results for two nuclear units proposed by the Public Service Electric & Gas Company of Newark, New Jersey.

PSE&G has proposed construction of two 1067-Mw BWR's at Hope Creek, to be completed in 1982 and 1984—the years that new capacity will be required to meet the company's projected 4.1% load growth rate. However, a study done for the State of New Jersey found that load growth reduction, to a rate of 2% or less, could be achieved with a series of technically feasible measures to increase energy use efficiency in New Jersey.[9] With a 2% growth rate, the need for Hope Creek 1 and 2 could be deferred until 1985 and 1990. Public Service's relative capital, fuel, and operating costs, assuming 2% load growth, for 1982

*Only early installation of nuclear units is covered since utilities do not generally advocate early installation of coal-fired units, as there is less running cost advantage over existing oil-fired units.

**Progress in improved energy technologies which might militate against early installation of facilities, using current technology, is deliberately excluded as being too speculative a factor.

***Note that the alternative to early installation is deferral of the entire plant construction process, rather than a stretch-out of construction over a longer time period. The latter is clearly uneconomical due to the interest which would accumulate during construction.

and 1984 installation of Hope Creek vs. 1985 and 1990 installation were compared for this study.*

It was found that deferral of Hope Creek installation to the later dates would reduce Public Service's revenue requirements between 1982-2000 by $145.1 million (1982 dollars), and by $57.6 million between 1982-2011—the projected period of operation of Hope Creek 1.**

The conclusion that savings accrue from deferring nuclear capacity initially appears questionable, since early installation of nuclear units reduces fuel costs and appears to reduce capital costs as well. Closer examination indicates, however, that capital cost escalation presently occurs at a slower rate than the cost of raising money to build plants (*i.e.*, Public Service projects an escalation rate of 5.52% per year if Hope Creek is deferred, versus an estimated cost of money of 10.78%). And so deferral reduces capital costs in constant dollars. The economics of deferral is, then, a trade-off between added fuel costs and reduced fixed charges. Notwithstanding the significant difference between nuclear and oil fuel costs, nuclear units, at low capacity factor, will not produce sufficient fuel cost savings to offset added fixed charges on a discounted cash flow basis. A capacity factor of 50% was used for Hope Creek 1 and 2.

Table 10.6 sets forth the year-by-year cost impact of the deferral of Hope Creek. In that table "Increase in Fixed Costs" presents the differential in fixed charges for Hope Creek as between (i) installation on Public Service's schedule and (ii) installation on the deferred schedule. In later years, there is a negative increase in fixed costs since deferral leads to higher capital costs (in current, non-deflated dollars).

"Increase in Production Costs" presents the increase in system running costs from operating existing units in lieu of deferred Hope Creek units (net of savings from avoiding O&M costs by deferring Hope Creek). Entries in this column are a function of projected nuclear fuel costs, nuclear O&M cost, replacement fuel and O&M cost, and Hope Creek capacity factor.***

Other columns provide a net sum of the fixed charges and production cost columns on an annual, discounted, and cumulative basis. The analysis is carried out over the first 30 years of station life, until nuclear units installed on Public Service's schedule would be retired.

Generally, early installation of nuclear units to substitute for oil-fired generation may create savings if nuclear capacity factors are 60%-65% or more. However, it is uneconomical at capacity factors below 50%-55% and indeterminate at 55%-60%. Precise assessment of early installation requires case-specific analysis of other cost parameters, accuracy of load forecasts, and potential of less tangible deferral benefits (such as the emergence of new energy

*For simplicity, CEP economic analysis of Public Service's proposed two-unit Atlantic nuclear station is omitted. See CEP's testimony before the New Jersey Public Utilities Commission.

**After 2011, savings from deferral would increase because installation of Hope Creek 1 in 1982 and 1984 (the non-deferral case) would require that the units be replaced in 2012 and 2014 at a significant capital cost (assuming 30-year life). The discounted cost of the new units installed in 2012 and 2014 would be greater than that of new units required in 2015 and 2020 under the deferral case.

***The values of these parameters are specified in the full CEP Testimony before the New Jersey Public Utilities Commission.

TABLE 10.6: ADDED COSTS FROM INSTALLATION OF HOPE CREEK 1 AND 2 IN 1982/84 RATHER THAN 1985/90

Year	Increase in Fixed Costs	Increase in Prodxn Costs	Increase in Annual Costs	Increase in Annual Costs Disc @10.78%	Cumulative Increase in Discounted Costs	
1982	147.4	−75.0	72.4	72.4	72.4	
1983	147.4	−76.8	70.6	63.7	136.1	
1984	311.4	−156.3	155.1	126.4	262.5	
1985	138.3	−79.5	58.8	43.3	305.8	
1986	138.3	−84.1	54.2	36.0	341.8	
1987	138.3	−89.0	49.3	29.5	371.3	
1988	138.3	−93.4	44.9	24.3	395.6	
1989	138.3	−98.9	39.4	19.2	414.8	
1990	−88.1	0	−88.1			
.	.	.	.			
.	.	.	.	→ −269.7 ↘		
.	.	.	.			
2000	−88.1	0	−88.1		145.1	Cumulative Added Costs Thru 2000
.	.	.	.			
.	.	.	.	→ −87.5 ↘		
.	.	.	.			
2011	−88.1	0	−88.1		57.6	Cumulative Added Costs Thru 2011

All Costs in $ Millions

supply technologies). CEP capacity factor trend equations suggest that early installation of 600-Mw nuclear units—especially PWR's—might be supported on cost grounds in some cases, but that early installation of the very large, 1150-Mw nuclear units is unlikely to reduce utility system costs.

REFERENCES

1. For prominent nuclear-coal comparative studies, see Bibliography.

2. *U.S. Atomic Energy Commission,* WASH-1345, "Power Plant Capital Costs: Current Trends and Sensitivity to Economic Parameters," October 1974. These ratios are consistent with coal to nuclear cost ratios in the Arthur D. Little and Ebasco Services reports (see Bibliography).

3. Testimony of Walter T. Schultheis, Director of Capacity Planning, Northeast Utilities, January 1976, Exhibit 6, in Hearings on the Economics of Nuclear Power before the Connecticut Public Utilities Control Authority.

4. See, for example, T.H. Lee, "Power Generation Economics," General Electric, January 22, 1976, submitted in the Connecticut PUCA hearings and circulated by the Atomic Industrial Forum.

5. Charles Komanoff and Raymond E. Gerson, "Economic Analysis of the Nuclear Expansion Program of Public Service Electric & Gas Company," CEP testimony before the New Jersey Public Utilities Commission, June 7, 1976, prepared for the New Jersey Department of Public Advocate, Appendix 2B2.

6. The 8 mill/kwh estimate is drawn from "1984 Steady-State Fuel Cycle Cost for PWR," Exhibit 10, in lecture by Michael Driscoll, Associate Professor of Nuclear Engineering, July 22, 1976, in M.I.T. Summer Course, "Principles of Nuclear Fuel and Power Management." It is consistent with the estimate by T.H. Lee, *op. cit.*, in the Connecticut PUCA hearings, with estimates by John P. Cagnetta, Director of Nuclear and Environmental Engineering, Northeast Utilities, in the same hearings, and with most current utility estimates of future nuclear fuel cycle costs. Some observers project significantly higher 1984 nuclear fuel costs due to higher costs for uranium and reprocessing and lower burnup. See, for example, Jim Harding, "Nuclear Fuel Burnup & Fuel Cycle Costs." California Energy Resources Conservation and Development Commission, memorandum, June 1976; or *Nucleonics Week,* "GESMO, Part 1, Out at Last: Reprocessing Gets Passing Marks—Just," September 2, 1976, Vol. 17, No. 36, p. 1.

7. See, for example, the Long Island Lighting Company, "Supplement 3, Revised," September 1974, submitted in Case 80003, LILCO Application before the New York State Board on Electric Generation Siting and the Environment, to construct two 1150-Mw nuclear units at Jamesport, Long Island.

8. CEP testimony in cases involving LILCO, Northeast Utilities and Public Service Electric & Gas (see Bibliography).

9. Dubin-Mindell-Bloome Associates, "A Study of Electrical Energy Usage in the Public Service Electric & Gas Company Service Territory," June 30, 1976, prepared for the New Jersey Department of Public Advocate.

Appendix A:
The CEP Capacity Factor Data Base

1. The Nuclear Data Base

The CEP data base for nuclear plants consists of all 38 U.S. nuclear units which entered commercial service between January 1, 1968, and December 31, 1974. These are 24 Pressurized Water Reactors (PWR), ranging in capacity from 450 Mw to 1050 Mw, and 14 Boiling Water Reactors (BWR), ranging from 514 Mw to 1098 Mw capacity. The data base includes operation for each unit beginning with the first calendar year in which the unit was in commercial service for the full year, through the end of 1975, or 66 unit-years for PWR's and 44 unit-years for BWR's. The units are operated by 23 different electric utilities.

Fifty-eight U.S. nuclear units were licensed to operate at the end of June, 1976 by the Nuclear Regulatory Commission, whereas the CEP data base includes only 38 units. The 20 units not in the CEP data base were excluded for the following five reasons:

Ten nuclear units entered commercial operation during 1975, but after New Year's Day, and therefore did not accumulate a full calendar-year of operation by the end of 1975 (Browns Ferry 2, Brunswick 2, Calvert Cliffs 1, Cook 1, Duane Arnold, Fitzpatrick, Hatch 1, Millstone 2, Rancho Seco, and Trojan).

One unit (St. Lucie 1) entered commercial operation during 1976.

Three units (Beaver Valley 1, Fort St. Vrain, and Indian Point 3) were in "power ascension" phase at the end of June, 1976. (Power ascension is a period of pre-commercial operation lasting several months in which the unit is operated at successively higher power levels to check the integrity of reactor equipment.)

Three units were rated at less than 100 Mw capacity (Big Rock Point 1, Humboldt Bay, and Lacrosse), and are therefore considerably different in design and operation from today's commercial reactors, which are in the 1000-Mw size class.

Three units in the 200-Mw size class (Dresden 1, Indian Point 1, and Yankee Rowe), are only half the size of the smallest unit in the CEP data base, and were installed six to eight years prior to the oldest unit in the data base. Accordingly, they are significantly different from current reactors.

The exclusion of these nuclear units generally results in a "conservative" formulation of the data base (that is, results in higher rather than lower nuclear capacity factors). For example, the ten units which entered commercial operation during 1975 operated during 1975 at an average capacity factor of only 40.6%.* This group, excluded from the data base, includes Browns Ferry 2 (4.5%) and Rancho Seco (20.6%), although it also includes other units with far higher capacity factors. Similarly, the three excluded nuclear units in the 200-Mw size class operated through 1975 at an average capacity factor of 54.1%, which is 5.2% less than the average capacity factor of the 38 units in the data base.

The exclusion of commercial operation prior to the first New Year's Day on which the unit was in commercial service is also conservative since nuclear units often operate at low capacity factors during such initial months due to mechanical problems and power level restrictions imposed in some cases by the Nuclear Regulatory Commission (NRC).**

Partial initial years of reactor operation were excluded in order to obtain an unambiguous "starting point" (New Year's Day) for measuring electrical generation to compute capacity factor. The reported date of commercial

*Computed as the average capacity factor in 1975 during all full months following the date of commercial operation, weighted by the number of such months. The capacity factor and number of months for each unit are: Browns Ferry 2, 4.5%, 10 months; Brunswick 2, 63.7%, 1 month; Calvert Cliffs 1, 76.0%, 7 months; Cook 1, 61.9%, 4 months; Duane Arnold, 49.5%, 11 months; Fitzpatrick 50.4%, 5 months; and Rancho Seco, 20.6%, 8 months. Hatch, Millstone 2, and Trojan were declared commercial near the end of December, 1975, and are therefore irrelevant to this computation.

**We have used the NRC definition of commercial service in the monthly NRC Gray Books ("Operating Units Status Reports"), which is "The date unit [is] declared by utility-owner to be available for the regular production of electricity; [is] usually related to satisfactory completion of qualification tests as specified in the purchase contract, and to accounting policies and practices of utility."

Selection by utilities of the date of commercial operation is thus somewhat discretionary. The time interval between date of initial "criticality"—the achievement of a self-sustaining fission reaction in the reactor—and the declared date of commercial operation, varies widely for the 38 nuclear units in the data base, with a mean of 5.7 months and a standard deviation of 4.9 months.

In an extreme case, Commonwealth Edison specified the commercial date for Dresden 2 in the Gray Books to be June 9, 1972, despite the fact that the unit went "critical" January 7, 1970—29 months earlier. The 1972 capacity factor of 44.8% for Dresden 2 (computed from net generation reported to the FPC) is *not* included in the CEP data base, which "picks up" Dresden 2 on January 1, 1973, based on the commercial date in the Gray Books.

In another instance, Florida Power and Light originally declared its new St. Lucie 1 plant to be in commercial service as of June 29, 1976, but subsequently re-designated the unit to power ascension status when the plant was shut two weeks later for the rest of the year to diagnose core power distribution problems (see July and August 1976 Gray Books).

For four other units, the year of initial commercial operation reported by the utility to the NRC is one year later than the year reported to the FPC (Quad Cities 1 and 2, Millstone 1, and Point Beach 2). In each case, correspondence with the utility established that the correct date is the date reported to the FPC. The unit-years included in the data base as a result, which are the first calendar years of operation for the four units, are generally high-capacity factor years (69%, 74%, 59%, and 69%, respectively) compared to other unit-years.

In some cases, utilities will declare a new unit commercial before the end of the year, for inclusion in the rate base. Fifteen of the 38 nuclear units have a December commercial operating date.

operation—which might be a logical starting point—is not feasible as a measuring index in many cases, since AEC and NRC records of annual generation generally do not begin counting electrical generation beginning on that date. Rather, AEC and NRC records frequently add electricity generated prior to commercial operation to aggregate generation data, occasionally causing an overstatement of capacity factor.* Finally, the data base is conservative omitting nuclear unit operation after December 31, 1975, since the average capacity factor of commercial nuclear units in the first six months of 1976 was only 52.2%, compared to 59.3% for the 1968-75 period in the CEP data base.

We have, however, compiled capacity factor data for the first six months of 1976 for the 38 units in the nuclear data base and the 10 additional nuclear units which entered commercial service in 1975. This 1976 data is discussed in Appendix D, but is not included in any of the figures or tables in the main text of this report, except where specifically noted.

The nuclear capacity factor figures in this study are computed on the basis of design electrical capacity ratings and annual net generation data, terms which are defined and discussed in Appendix B. The design ratings were taken from the NRC Gray Book for January 1976. Annual net generation data for 1975 and 1974 was obtained from the January 1976 and January 1975 Gray Books, respectively. Generation data for years prior to 1974 was taken from the annual FPC report, "Steam-Electric Plant Construction Cost and Annual Production Expenses" and, in several cases, from private communications with utilities. Data for the first six months of 1976 was taken from the July 1976 Gray Book.

Below we have listed the capacity factor data and related parameters used in this report, in three sections:

(1) the CEP nuclear data base, totalling 110 years of operation for 38 reactors during 1968-1975.

(2) 1976 six-month data, totalling 24 unit-years for 48 reactors.

(3) 1961-1976 operation for three 200-Mw class reactors not in the data base—Dresden 1, Indian Point 1, and Yankee Rowe (43½ unit-years).

2. The Coal-Fired Unit Data Base

The CEP data base for coal-fired units consists of 1641 unit-years of operation by 250 coal-fired generating units, operated by 86 electric utilities. This sample includes the years of operation 1961-1973 for all coal-fired units 100 Mw and larger, which entered commercial service in the U.S. during 1960-1972.** As was done for nuclear units, the coal data base excludes initial partial years of operation and counts a unit's first calendar year of operation (beginning on the first New Year's Day on which the unit was in commercial service) as the first year of operation.

*Accurate monthly generation data has been readily available since mid-1974 from the NRC Gray Books. Monthly data prior to that time is available in theory from regular utility reports on file in the NRC Public Document Room. However, such data was absent from the files for over half the nuclear units in the data base when a CEP researcher attempted to record the data in late 1975.

**The only unit-year omitted from the sample was the year of operation 1961 for Appalachian Power Company's Sporn #5 unit, for which net generation data was unavailable from either the company or government sources.

As with nuclear units, the exclusion of initial partial years of operation for coal units was dictated by difficulties in collecting such data, compounded in this case by the absence of partial year reporting by utilities to the federal government. Exclusion of initial partial years tends to increase capacity factor averages, due to the break-in period generally involved when new coal units (or nuclear units) are brought into commercial service.

The CEP coal data base is the first complete record of capacity factors to be compiled for modern U.S. coal-fired units. The only comparable data base, that of the Equipment Availability Task Force of the Edison Electric Institute, is not comprehensive since utilities report data to EEI on a voluntary basis and sometimes do not file reports for certain unit-years. It is based on ratings of generation and capacity which are not suitable for economic analysis (see discussion in Appendix B).*

The coal-fired generating units in the CEP data base were identified by consulting the annual FPC report, "Steam-Electric Plant Construction Cost And Annual Production Expenses," which lists each year's new generating units, by fuel type, for the years 1960-1972. For units with multiple fuel firing, all unit-years were included in which 60% or more of the fuel input was provided by coal.

Net generation data was obtained for about 1400 of the 1641 unit-years from Schedule 432b of each utility's annual "Form 1" Report to the FPC, for the years 1961-1973. Schedule 432b lists a utility's most efficient generating units for that year and includes the net generation for each unit listed.**

The unit-years listed in Schedule 432b, taken alone, would have been an unrepresentative sample for a capacity factor analysis since Schedule 432b specifically excludes units with less than a 50% capacity factor for a given year. Therefore, we obtained generation data for the several hundred unit-years not available from Schedule 432b by letter from the 59 utilities which operate the relevant units. Together, the data from Schedule 432b and the utilities constitutes a complete record of net electrical generation by U.S. coal-fired units, within the size and date constraints referred to earlier.***

As we discuss in Appendix B, the "maximum nameplate generator rating" was chosen as the measure of Mw capacity for coal-fired units in this study. The nameplate ratings of coal units in the data base were compiled from the

*In addition, the EEI data base is not open to scrutiny on a unit-by-unit basis, although the EEI Task Force publishes an annual report summarizing its statistical analysis of the data base and also performs special data analyses on an aggregate basis for applications meeting EEI criteria.

**The FPC Steam-Electric Plant Books, referred to above, list the net generation of each generating *station*, but do not provide generation data for separate *units*, except for single-unit stations. Since most coal-fired units are sited at stations which include other units, the FPC Steam-Electric Plant Books provide generation data for only a fraction of U.S. coal-fired units.

***The coal unit data base in the study "lags" the nuclear data base by two years, since it includes only units installed through 1972, and years of operation through 1973. The earlier cut-off point for coal-units was dictated by the availability of utility data from the FPC at the time the sample data was collected (early 1975). The exclusion of subsequent experience with coal-fired units, although regrettable, eliminates the problem of adjusting for extensive load-following by some coal-fired units in 1974-1975, resulting from sharply reduced load growth in the wake of the Arab oil embargo.

capacity ratings given in the annual FPC Steam Plant Cost Books.* In the dozen or so cases in which units subsequently were re-rated by utilities because of changes in the definition of capacity rather than physical modifications of the units, we have used the later capacity rating, which was higher than the initial rating in most cases.

Coal quality variables (sulfur, ash, and heat content) for the years 1969-71 were compiled from the annual FPC report, Steam-Electric Plant Air and Water Quality Control Data for those years. Coal quality data for 1972 and 1973 was compiled from data sheets for the reports for those years in the FPC Section of Fuel and Environmental Analysis. Coal quality data for 1969 was applied to pre-1969 operation on the assumption that relatively little fuel switching or other major variation in coal quality occurred at U.S. plants prior to 1970.**

Of the 1641 coal unit-years in the data base, complete coal quality data was unavailable for 14 unit years. (These were primarily units which converted to oil before 1969, so that coal quality data prior to 1969 was unavailable.) Capacity factor averages for coal units in this study are based on the 1641 unit-years, while the capacity performance regression equations are based on 1627 unit-years. The capacity factors of the 14 unit-years excluded from the regression equation are not substantially different from the sample as a whole.

Below we have listed the station name, operating utility, lifetime capacity factor and capacity performance, and key operating parameters of the 250 coal-fired units in the CEP data base. Performance variables are also listed for each of the 86 operating utilities. In addition, we have lised the annual capacity factor and performance of each unit of 400 Mw capacity or larger. This sample comprises 93 units operating for 381 unit-years. A computer tape and listing of the entire 1641 unit-years in the data base is available by special request from CEP.

*Since the FPC Steam Plant Cost Books report statistics for stations rather than individual units, unit capacity was compiled by comparing a station's installed generating capacity, as listed in the FPC Steam Plant Cost Books, for the year before and the year during which the new unit was installed.

**This assumption is considered accurate by Alexander Gakner, Head, Section of Fuel and Environmental Analysis, FPC (telephone communication, June, 1975). Utilities were first required to report fuel quality to the FPC for 1969.

CEP NUCLEAR DATA BASE, ANNUAL CAPACITY DATA

Unit	Utility	Mfg	MW Capacity FPC	NRC	MDC	Comm. Date	Year	GWH	CF%
1. Arkansas 1	Arkansas P&L	BW	897	850	836	12/19/74	1975	4880	65.5
2. Pilgrim 1	Boston Edison	GE	655	670	670	12/-/74	1973	4074	69.4
							1974	1973	33.6
							1975	2587	44.1
3. Robinson 2	Carolina P&L	W	769	707	665	3/7/71	1972	4829	78.0
							1973	3765	60.8
							1974	4813	77.7
							1975	4171	67.3
5. Dresden 2	Commonwealth Ed	GE	828	809	800	6/9/72	1973	5015	70.8
							1974	3380	47.7
							1975	2966	41.9
6. Dresden 3	Commonwealth Ed	GE	828	809	800	11/16/71	1972	4933	69.6
							1973	3694	52.1
							1974	3200	45.2
							1975	2190	30.9
7. Quad Cities 1	Commonwealth Ed	GE	828	809	800	2/18/73	1973	4909	69.3
							1974	3563	50.3
							1975	4271	60.3
8. Quad Cities 2	Commonwealth Ed	GE	828	809	800	3/10/73	1973	5209	73.5
							1974	4470	63.1
							1975	2475	34.9

Unit	Utility	Mfg	MW Capacity FPC	NRC	MDC	Comm. Date	Year	GWH	CF%
9. Zion 1	Commonwealth Ed	W	1098	1050	880	12/31/73	1974	3478	37.8
							1975	4909	53.4
10. Zion 2	Commonwealth Ed	W	1098	1050	880	9/17/74	1975	4829	52.5
12. Indian Point 2	Consolidated Ed	W	873	873	864	8/-/73	1974	3324	43.5
							1975	4885	63.9
13. Palisades	Consumers Power	CE	812	821	684	12/31/71	1972	1765	24.5
							1973	2410	33.5
							1974	78	1.1
							1975	2428	33.8
14. Oconee 1	Duke Power	BW	887	886	871	7/15/73	1974	3999	51.5
							1975	5286	68.1
15. Oconee 2	Duke Power	BW	887	886	871	7/11/74	1975	4968	64.0
16. Oconee 3	Duke Power	BW	887	886	871	12/16/74	1975	5037	64.9
17. Turkey Point 3	Florida P&L	W	760	745	666	12/14/72	1973	3306	50.7
							1974	3624	55.5
							1975	4375	67.0
18. Turkey Point 4	Florida P&L	W	760	745	666	9/7/73	1974	4292	65.8
							1975	3990	61.1
19. Oyster Creek	Jersey Central	GE	550	650	620	12/-/69	1970	3454	60.7
							1971	3825	67.2
							1972	4356	76.5

Unit	Utility	MFG	MW Capacity FPC	NRC	MDC	Comm. Date	Year	GWH	CF%
							1973	3585	63.0
							1974	3674	64.5
							1975	3146	55.3
20. 3 Mile Island 1	Metropolitan Ed	BW	871	819	792	9/2/74	1975	5542	77.2
21. Cooper Station	Nebraska PPD	GE	801	778	764	7/1/74	1975	3854	56.5
22. Maine Yankee	Me. Yankee Atom.	CE	830	790	790	12/28/72	1973	3351	48.4
							1974	3574	51.6
							1975	4502	65.1
23. Vermont Yankee	Vt. Yankee Nuc.	GE	514	514	514	11/30/72	1973	1814	40.3
							1974	2483	55.1
							1975	3561	79.1
25. Nine Mile Pt. 1	Niagara Mohawk	GE	620	610	610	12/-/69	1970	1687	31.6
							1971	2937	55.0
							1972	3242	60.7
							1973	3494	65.4
							1974	3297	61.7
							1975	3045	57.0
26. Millstone 1	Northeast Utils	GE	662	690	650	3/-/71	1971	3580	59.2
							1972	3176	52.5
							1973	1880	31.1
							1974	3604	59.6
							1975	3897	64.5
27. Conn Yankee	Northeast Utils	W	600	575	575	1/1/68	1968	2995	59.5

Unit	Utility	Mfg	MW Capacity FPC	NRC	MDC	Comm. Date	Year	GWH	CF%
							1969	3639	72.2
							1970	3538	70.2
							1971	4187	83.1
							1972	4300	85.4
							1973	2425	48.1
							1974	4351	86.4
							1975	4121	81.8
28. Monticello	No. States Power	GE	569	545	538	7/4/71	1972	3559	74.5
							1973	3271	68.5
							1974	2924	61.2
							1975	2879	60.3
29. Prairie Island 1	No. States Power	W	593	530	520	12/5/73	1974	1433	30.9
							1975	3694	79.6
30. Prairie Island 2	No. States Power	W	593	530	520	12/21/74	1975	3176	68.4
31. Fort Calhoun	Omaha PPD	CE	481	457	457	6/20/74	1975	2081	52.0
32. Peach Bottom 2	Phila. Electric	GE	1152	1065	1051	5/5/74	1975	5082	54.5
33. Peach Bottom 3	Phila. Electric	GE	1152	1065	1035	12/23/74	1975	5282	56.6
34. Ginna	Rochester G&E	W	517	490	470	3/-/70	1971	2705	63.0
							1972	2356	54.9
							1973	3399	79.2
							1974	2097	48.9
							1975	3041	70.8
35. San Onofre 1	So. Calif Edison	W	450	450	430	1/1/68	1968	1262	32.0

Unit	Utility	Mfg	MW Capacity FPC	NRC	MDC	Comm.Date	Year	GWH	CF%
							1969	2607	66.1
							1970	3059	77.6
							1971	3303	83.8
							1972	2812	71.3
							1973	2273	57.7
							1974	3145	79.8
							1975	3245	82.3
36. Browns Ferry 1	TVA	GE	1152	1098	1065	8/1/74	1975	1348	14.0
37. Surry 1	Vepco	W	848	823	788	12/22/72	1973	3480	48.3
							1974	3318	46.0
							1975	3917	54.3
38. Surry 2	Vepco	W	848	823	788	5/1/73	1974	2635	36.5
							1975	5053	70.1
39. Point Beach 1	Wisc Elec Power	W	524	497	490	12/21/70	1971	3274	75.2
							1972	2943	67.6
							1973	2742	63.0
							1974	3142	72.2
							1975	2922	67.1
40. Point Beach 2	Wisc Elec Power	W	524	497	490	4/20/73	1973	2991	68.7
							1974	3178	73.0
							1975	3741	85.9
41. Kewaunee	Wisc Pub Serv	W	535	560	535	6/-/74	1975	3341	68.1

ANNUAL CAPACITY DATA FOR THREE PRE-COMMERCIAL NUCLEAR UNITS

			MW Capacity						
Unit	Utility	MFG	FPC	NRC	MDC	Comm. Date	Year	GWH	CF%
4. Dresden 1	Commonwealth Ed	GE	209	200	200	7/4/60	1961	528	30.1
							1962	1193	68.1
							1963	934	53.3
							1964	980	55.9
							1965	962	54.9
							1966	1407	80.3
							1967	807	46.1
							1968	916	52.3
							1969	826	47.1
							1970	1361	77.7
							1971	603	34.4
							1972	1099	62.7
							1973	560	32.0
							1974	353	20.1
							1975	697	39.8
11. Indian Point 1	Consolidated Ed	BW	275	265	257	10/-/62	1963	849	36.6
							1964	551	23.7
							1965	1037	44.7
							1966	1124	48.4
							1967	1622	69.9
							1968	1511	65.1
							1969	1674	72.1
							1970	328	14.1
							1971	1228	52.9
							1972	1128	48.6
							1973	0	0.0
							1974	1227	52.9

Unit	Utility	Mfg	MW Capacity FPC	NRC	MDC	Comm. Date	Year	GWH	CF%
							1975	0	0.0
24. Yankee Rowe	New Engl. Elec	W	185	175	175	7/-/61	1962	690	45.0
							1963	942	61.4
							1964	1182	77.1
							1965	966	63.0
							1966	1295	84.5
							1967	1273	83.0
							1968	1218	79.5
							1969	1143	74.6
							1970	1209	78.9
							1971	1435	93.6
							1972	648	42.3
							1973	1046	68.2
							1974	911	59.4
							1975	1193	77.8

1976 6-MONTH DATA FOR NUCLEAR UNITS
(January 1, 1976-June 30, 1976)

Units in CEP Data Base

Unit	Utility	Mfg	MW	Comm. Date	GWH	CF
1. Arkansas 1	Arkansas P&L	BW	850	12/19/74	1178	31.7
2. Pilgrim 1	Boston Edison	GE	670	12/-/74	349	11.9
3. Robinson 2	Carolina P&L	W	707	3/7/71	2800	90.7
5. Dresden 2	Commonwealth Ed	GE	809	6/9/72	1523	43.1
6. Dresden 3	Commonwealth Ed	GE	809	11/16/71	2286	64.7
7. Quad Cities 1	Commonwealth Ed	GE	809	2/18/73	1157	32.7
8. Quad Cities 2	Commonwealth Ed	GE	809	3/10/73	2658	75.2
9. Zion 1	Commonwealth Ed	W	1050	12/31/73	1319	28.8
10. Zion 2	Commonwealth Ed	W	1050	9/17/74	1831	39.9
12. Indian Point 2	Consolidated Ed	W	873	8/-/73	1507	39.5
13. Palisades	Consumers Power	CE	821	12/31/71	606	16.9
14. Oconee 1	Duke Power	BW	886	7/15/73	1413	36.5
15. Oconee 2	Duke Power	BW	886	7/11/74	1595	41.2
16. Oconee 3	Duke Power	BW	886	12/16/74	2720	70.3
17. Turkey Point 3	Florida P&L	W	745	12/14/72	2412	74.1
18. Turkey Point 4	Florida P&L	W	745	9/7/73	1736	53.4
19. Oyster Creek	Jersey Central	GE	650	12/-/69	1367	48.2
20. 3 Mile Island 1	Metropolitan Ed	BW	819	9/2/74	1517	42.4
21. Cooper Station	Nebraska PPD	GE	778	7/1/74	2008	59.1
22. Maine Yankee	Me. Yankee Atom.	CE	790	12/28/72	3041	88.1

Unit	Utility	Mfg	MW	Comm. Date	GWH	CF
23. Vermont Yankee	Vt. Yankee Nuc.	GE	514	11/30/72	1672	74.5
25. Nine Mile Pt. 1	Niagara Mohawk	GE	610	12/-/69	1911	71.7
26. Millstone 1	Northeast Utils	GE	690	3/-/71	2263	75.1
27. Conn Yankee	Northeast Utils	W	575	1/1/68	1820	72.5
28. Monticello	No. States Power	GE	545	7/4/71	2008	84.4
29. Prairie Island 1	No. States Power	W	530	12/5/73	1199	51.8
30. Prairie Island 2	No. States Power	W	530	12/21/74	1396	60.3
31. Fort Calhoun	Omaha PPD	CE	457	6/20/74	1294	64.8
32. Peach Bottom 2	Phila. Electric	GE	1065	5/5/74	1966	42.3
33. Peach Bottom 3	Phila. Electric	GE	1065	12/23/74	2444	52.6
34. Ginna	Rochester G&E	W	490	3/-/70	794	37.1
35. San Onofre 1	So. Calif. Edison	W	450	1/1/68	1741	88.6
36. Browns Ferry 1	TVA	W	1098	8/1/74	0	0.0
37. Surry 1	VEPCO	GE	823	12/22/72	2875	80.0
38. Surry 2	VEPCO	W	823	5/1/73	1922	53.5
39. Point Beach 1	Wisc. Elec. Power	W	497	12/21/70	1946	89.7
40. Point Beach 2	Wisc. Elec. Power	W	497	4/20/73	1644	75.7
41. Kewaunee	Wisc. Pub Serv	W	560	6/-/74	1192	48.7
Units Entering Commercial Service in 1975						
42. Browns Ferry 2	TVA	GE	1098	3/1/75	0	0.0
43. Brunswick	Carolina P & L	GE	821	11/3/75	1117	31.2
44. Calvert Cliffs 1	Baltimore G&E	CE	845	5/8/75	3165	85.8
45. Cook 1	Ind. & Mich. Elec.	W	1090	8/27/75	2999	63.0
46. Duane Arnold	Iowa Elec L&P	GE	538	2/1/75	866	36.9
47. Fitzpatrick	NYS Power Auth	GE	821	7/28/75	1661	46.3
48. Hatch 1	Georgia Power	GE	786	12/31/75	1750	51.0

Unit	Utility	Mfg	MW	Comm. Date	GWH	CF
49. Millstone 2	Northeast Utils	CE	828	12/26/75	1917	53.0
50. Rancho Seco	Sacramento M.U.D.	BW	913	4/17/75	716	18.0
51. Trojan	Portland Gen Elec	W	1130	12/24/75	512	10.4
Pre-Commercial Units						
4. Dresden 1	Commonwealth Ed	GE	200	7/4/60	474	54.3
11. Indian Point 1	Consolidated Ed	BW	265	10/-/62	0	0.0
24. Yankee Rowe	New Engl. Elec	W	175	7/-/61	531	69.5

UTILITY AVERAGE COAL CAPACITY STATISTICS

Utility	# of Units In Data Base	# of Unit-Years	Avg. C.F.	Avg. C.P.	Avg. MW Cpcty	Avg. Sulfur Content	% Unit-Years Supercritical
1. Alabama Power	5	22	61.0	67.6	375	2.1	14
2. Appalachian Power	4	27	77.3	86.4	417	1.4	56
3. Arizona Public Service	6	47	75.3	85.7	266	0.6	15
4. Associated Elec Coop (Mo.)	3	12	69.6	78.8	247	—	0
5. Atlantic City Electric	2	16	73.9	85.3	148	2.7	0
6. Baltimore Gas & Elec	3	23	70.8	79.5	246	2.2	30
7. Basin Elec Coop (N.D.)	1	8	66.9	73.8	240	0.7	0
8. Big Rivers Rural Elec (Ky.)	3	9	71.7	82.8	171	3.8	0
9. Carolina P & L	6	47	67.2	74.4	296	1.2	0
10. Central Hudson G&E	1	2	65.0	71.8	239	1.8	0
11. Central Illinois Light	3	16	65.4	74.4	192	2.8	0
12. Central Illinois P S	3	22	57.0	62.5	311	3.7	0
13. Cincinnati G&E	3	28	74.0	83.9	238	3.1	0
14. Cleveland Elec Illuminating	3	15	66.6	73.0	369	3.2	27
15. Colorado-Ute Elec Assoc	1	8	87.5	101.1	163	4.6	0
16. Columbus & Southern Ohio Elec	4	16	68.2	77.3	302	3.7	31
17. Commonwealth Edison (Ill.)	9	65	56.8	61.6	527	3.2	0
18. Connecticut L&P	2	17	76.1	87.9	163	2.2	0

Utility	#of Units In Data Base	#of Unit-Years	Avg. C.F.	Avg. C.P.	Avg. MW Cpcty	Avg. Sulfur Content	% Unit-Years Supercritical
19. Consolidated Edison (N.Y.)	3	20	47.0	50.9	483	0.9	0
20. Consumers Power Power (Mich.)	3	29	76.3	84.0	290	2.9	21
21. Dairyland Power Coop (Wisc.)	1	4	54.3	58.8	346	3.7	100
22. Dallas P&L	2	2	64.4	70.8	593	0.6	100
23. Delmarva P&L	2	8	67.4	77.9	177	2.0	0
24. Detroit Edison	4	23	72.4	79.0	467	2.8	9
25. Duke Power (N.C./S.C.)	7	48	82.7	90.3	359	1.0	15
26. Duquesne Light (Pa.)	2	16	78.1	89.6	256	2.0	0
27. East Kentucky Power Coop	2	12	61.5	70.1	153	2.7	0
28. Empire District Elec (Mo.)	1	3	66.4	73.2	213	4.7	0
29. Florida Power Corp	2	5	50.0	56.4	474	3.4	0
30. Georgia Power	10	57	58.9	65.3	364	1.4	21
31. Gulf Power (Fla.)	2	14	66.3	76.6	167	2.9	0
32. Hartford Elec Light	1	3	67.8	74.8	239	—	0
33. Holyoke Water Power (Mass.)	1	9	91.0	105.1	136	2.2	0
34. Illinois Power	2	12	64.8	70.1	447	3.2	0
35. Indiana & Michigan Elec	2	22	58.9	66.4	530	3.4	91
36. Indiana Statewide Rural	2	8	65.1	75.2	117	2.7	0

Utility	# of Units In Data Base	# of Unit-Years	Avg. C.F.	Avg. C.P.	Avg. MW Cpcty	Avg. Sulfur Content	% Unit-Years Supercritical
37. Indianapolis P&L	3	22	68.3	77.8	217	3.5	0
38. Interstate Power (Iowa)	1	6	59.8	66.0	218	3.0	0
39. Iowa Elec L&P	1	5	53.8	62.1	149	2.6	0
40. Iowa P S	1	1	51.3	55.6	349	0.6	0
41. Iowa Southern Utilities	1	5	54.1	59.7	212	2.7	0
42. Kansas City P&L	2	22	66.1	76.3	188	5.9	0
43. Kentucky Power	2	15	72.6	79.7	424	1.0	27
44. Kentucky Utilities	2	12	50.1	57.8	222	2.6	0
45. Louisville G&E	4	23	63.6	72.0	204	3.4	0
46. Metropolitan Edison (Pa.)	1	11	73.5	81.1	255	2.3	0
47. Minnkota Power Coop (N.D.)	1	3	82.8	91.3	235	0.7	0
48. Mississippi Power	1	5	50.1	55.3	299	2.6	0
49. Missouri Public Service	1	4	35.9	41.1	419	3.7	100
50. Monogahela Power (W. Va.)	4	25	76.9	87.0	367	3.7	48
51. Montana Power	1	5	48.4	56.0	173	0.7	0
52. Nevada Power	2	13	78.6	90.8	114	0.6	0
53. New England Power	2	10	85.0	93.8	241	1.0	0
54. Northern Indiana P S	3	19	65.2	75.2	242	3.7	26
55. Northern States Power (Minn.)	3	27	74.1	83.8	269	3.0	19
56. Ohio Edison	6	48	74.9	85.5	256	2.8	13
57. Ohio Power	5	21	66.4	71.5	653	3.7	100

Utility	# of Units In Data Base	# of Unit-Years	Avg. C.F.	Avg. C.P.	Avg. MW Cpcty	Avg. Sulfur Content	% Unit-Years supercritical
58. Omaha Public Power District	2	15	68.4	78.0	163	2.6	0
59. Orange & Rockland (N.Y.)	1	2	61.9	71.6	179	1.7	0
60. Pacific P&L (Wyo.)	3	22	61.6	69.9	160	0.5	0
61. Pennsylvania Elec*	3	21	66.6	75.6	368	2.7	38
62. Pennsylvania P&L	8	41	59.4	65.1	647	2.4	51
63. Pennsylvania Power Co.	1	9	70.6	81.6	133	3.4	0
64. Philadelphia Electric	2	26	71.2	77.1	354	2.3	100
65. Potomac Electric	4	42	66.5	75.2	268	2.2	43
66. Public Service of Colorado	3	25	66.5	75.9	221	0.6	0
67. Public Service of Indiana	4	21	65.2	73.8	279	3.1	0
68. Public Service New Hampshire	2	18	72.6	82.6	178	2.4	0
69. Public Service E&G (N.J.)	3	26	67.1	72.9	331	1.9	4
70. Savannah E&P	1	1	63.0	72.8	104	2.5	0
71. South Carolina E&G	5	31	58.0	66.0	192	1.2	16
72. South Carolina PS Authority	2	7	58.1	67.1	173	1.1	0
73. Southern California Edison	2	4	47.7	51.4	818	0.4	100
74. Southern Elec Gen (Alà.)	4	49	80.5	88.9	266	1.0	0
75. Tampa Electric	5	39	51.5	58.9	250	3.6	0
76. Tennessee Valley Authority	7	57	61.0	66.4	683	3.4	16

*Penn Elec figures do not include 4 Penn P&L units constructed and operated by Penn Elec.

Utility	# of Units In Data Base	# of Unit-Years	Avg. C.F.	Avg. C.P.	Avg. MW Cpcty	Avg. Sulfur Content	% Unit-Years Supercritical
77. Toledo Edison	2	15	75.8	86.4	167	2.1	0
78. Union Electric (Mo.)	6	29	54.4	59.2	486	2.9	38
79. United Illuminating (Conn.)	1	5	66.6	76.9	180	—	0
80. United Power Assoc (N.D.)	1	6	68.0	78.5	172	0.8	0
81. Utah P&L	3	17	64.9	73.5	198	0.5	0
82. Virginia Elec & Power	7	47	62.6	69.6	348	1.4	0
83. West Penn Power	4	19	67.5	74.4	430	2.8	47
84. Wisconsin Elec Power	3	26	59.4	64.8	300	2.2	0
85. Wisconsin P&L	2	15	67.7	76.9	177	3.0	0
86. Wisconsin Public Service	1	9	85.6	98.8	125	2.5	0

LIFETIME COAL CAPACITY STATISTICS BY UNIT

Unit	Utility* Number	Years In Data Base	Capacity Factor	Capacity Performance	Mw	Avg. Sulfur Content	Year Installed
1. Barry 4	1	4	50.9	58.3	404	2.5	1969
2. Barry 5 †	1	2	43.3	46.6	789	2.5	1971
3. Gorgas 10 †	1	1	57.9	62.4	789	1.2	1972
4. Greene County 1	1	8	66.0	72.9	299	2.1	1965
5. Greene County 2	1	7	66.7	73.6	269	2.1	1966
6. Amos 1 †	2	2	79.8	86.0	816	1.0	1971
7. Amos 2 †	2	1	79.8	86.0	816	1.0	1972
8. Clinch River 3	2	12	85.6	94.5	238	0.6	1961
9. Sporn 5 †	2	12	68.5	78.4	496	2.3	1960
10. Cholla 1	3	11	83.2	96.2	114	0.5	1962
11. Four Corners 1	3	10	80.6	93.1	173	0.6	1963
12. Four Corners 2	3	10	77.1	89.0	173	0.6	1963
13. Four Corners 3	3	9	67.5	74.5	230	0.6	1964
14. Four Corners 4 †	3	4	62.7	67.6	818	0.7	1969
15. Four Corners 5 †	3	3	63.4	68.3	818	0.7	1970
16. Thomas Hill 1	4	7	69.6	80.5	175	4.2	1966
17. Thomas Hill 2	4	4	74.6	81.3	272	4.3	1969
18. New Madrid 1	4	1	49.1	52.9	650	—	1972
19. England 1	5	9	71.9	82.1	136	2.8	1962
20. England 2	5	7	77.4	89.5	163	2.6	1964
21. Crane 1	6	10	77.6	89.7	190	2.4	1961
22. Crane 2	6	6	66.6	73.5	209	2.4	1963

*Derived from table of Utility Average Coal Capacity statistics.

Unit	Utility Number	Years In Data Base	Capacity Factor	Capacity Performance	Mw	Avg. Sulfur Content	Year Installed
23. Wagner 3 †	6	7	64.7	70.2	359	1.7	1966
24. Leland Olds 1	7	8	66.9	73.8	240	0.7	1966
25. Coleman 1	8	4	69.3	80.0	170	3.9	1969
26. Coleman 2	8	3	70.2	81.1	170	3.8	1970
27. Coleman 3	8	2	78.9	91.1	173	3.8	1971
28. Asheville 1	9	9	70.2	77.5	207	1.2	1964
29. Asheville 2	9	2	66.6	73.5	207	1.4	1971
30. Fitzhugh Lee 3	9	11	72.8	80.4	252	1.0	1962
31. H.B. Robinson 1	9	13	58.9	65.0	207	1.2	1960
32. Roxboro 1	9	7	67.6	77.4	411	1.4	1966
33. Roxboro 2	9	5	70.5	76.0	657	1.4	1968
34. Danskammer 4	10	2	65.0	71.8	239	1.8	1967
35. Edwards 1	11	10	67.9	78.4	136	2.8	1960
36. Edwards 2	11	5	62.5	69.0	268	2.7	1968
37. Edwards 3	11	1	55.7	60.4	364	2.8	1972
38. Coffeen 1	12	8	51.2	55.5	389	4.5	1965
39. Coffeen 2	12	1	42.3	45.6	617	4.4	1972
40. Meredosia 3	12	13	61.7	68.0	239	3.2	1960
41. Beckjord 5	13	11	75.0	82.8	245	2.8	1962
42. Beckjord 6	13	4	51.1	58.5	461	2.8	1969
43. Miami Fort 6	13	13	80.1	92.6	163	3.4	1960
44. Avon 9 †	14	3	65.0	70.1	680	2.8	1970
45. Eastlake 5 †	14	1	66.2	71.3	680	3.5	1972
46. Lake Shore 18	14	11	67.0	74.0	256	3.2	1962

Unit	Utility Number	Years In Data Base	Capacity Factor	Capacity Performance	Mw	Avg. Sulfur Content	Year Installed
47. Hayden 1	15	8	87.5	101.1	163	0.5	1965
48. Conseville 3	16	11	69.9	80.8	162	4.7	1962
49. Stuart 2 †	16	2	62.0	66.8	610	1.7	1971
50. Stuart 1 †	16	2	61.1	65.8	610	1.6	1970
51. Stuart 3 †	16	1	76.2	82.1	610	1.7	1972
52. Crawford 8	17	10	68.0	73.7	358	2.8	1961
53. Joliet 7	17	8	51.8	55.8	660	3.4	1965
54. Joliet 8	17	7	53.1	57.3	660	3.3	1966
55. Kinkaid 1	17	6	42.2	45.5	660	4.1	1967
56. Kinkaid 2	17	5	46.5	50.1	660	4.1	1968
57. Powerton 5	17	1	51.6	55.6	893	3.6	1972
58. State Line 4	17	7	65.8	71.3	389	2.9	1962
59. Waukegan 8	17	11	64.4	69.8	355	2.7	1962
60. Will County 4	17	10	52.1	57.4	598	2.9	1963
61. Norwalk Harbor 1	18	9	73.6	85.0	163	2.2	1960
62. Norwalk Harbor 2	18	8	78.9	91.1	163	2.2	1963
63. Astoria 4	19	9	41.1	44.5	387	0.9	1961
64. Astoria 5	19	8	55.1	59.7	387	0.9	1962
65. Ravenswood 3	19	3	43.0	46.3	1028	0.8	1965
66. Campbell 1	20	11	76.0	83.9	265	3.3	1962
67. Campbell 2 †	20	6	63.7	69.0	385	3.4	1967
68. Karn 2	20	12	82.9	91.5	265	2.3	1961
69. New Genoa 3 †	21	4	54.3	58.8	346	3.7	1969
70. Big Brown 1 †	22	1	65.7	72.3	593	0.6	1971

Unit	Utility Number	Years In Data Base	Capacity Factor	Capacity Performance	Mw	Avg. Sulfur Content	Year Installed
71. Big Brown 2 †	22	1	63.0	69.4	593	0.6	1972
72. Edgemoor 4	23	5	70.8	81.8	177	2.1	1966
73. Indian River 3	23	3	61.8	71.4	177	1.8	1970
74. Monroe 1 †	24	2	52.3	56.4	817	2.8	1971
75. St. Clair 6	24	12	74.9	81.2	353	3.0	1961
76. St. Clair 7	24	4	70.1	77.1	545	2.5	1969
77. Trenton Channel 9	24	5	76.3	84.0	536	2.6	1968
78. Allen 4	25	13	75.3	83.1	275	1.0	1960
79. Allen 5	25	12	79.4	87.6	275	1.0	1961
80. Cliffside 5	25	1	69.4	76.4	571	1.3	1972
81. Marshall 1	25	8	95.6	103.6	350	1.0	1965
82. Marshall 2	25	7	96.7	104.9	350	1.1	1966
83. Marshall 3 †	25	4	73.5	79.2	650	1.0	1969
84. Marshall 4 †	25	3	76.8	82.8	650	1.0	1970
85. Cheswick 1	26	3	64.1	70.6	565	2.2	1970
86. Elrama 4	26	13	81.3	94.0	185	1.9	1960
87. Cooper 1	27	8	62.5	72.2	114	2.7	1965
88. Cooper 2	27	4	59.7	65.8	230	2.6	1969
89. Asbury 1	28	3	66.4	73.2	213	4.7	1970
90. Crystal River 1	29	3	53.5	61.3	440	3.3	1966
91. Crystal River 2	29	2	44.6	49.1	524	3.5	1969
92. Bowen 1 †	30	2	43.6	47.0	806	2.8	1971
93. Bowen 2 †	30	1	53.1	57.3	806	3.1	1972
94. Hammond 4	30	3	39.7	43.6	578	2.8	1970

Unit	Utility Number	Years In Data Base	Capacity Factor	Capacity Performance	Mw	Avg. Sulfur Content	Year Installed
95. Harllee Branch 1	30	8	59.8	66.0	299	1.3	1965
96. Harllee Branch 2	30	6	56.7	61.5	359	1.3	1967
97. Harllee Branch 3 †	30	5	52.0	57.2	544	1.3	1968
98. Harllee Branch 4 †	30	4	43.0	47.3	544	1.2	1969
99. McDonough 1	30	10	67.1	74.1	299	1.1	1963
100. McDonough 2	30	9	69.2	76.4	299	1.1	1964
101. Mitchell 3	30	9	61.5	71.1	163	1.3	1964
102. Lansing Smith 1	31	8	69.4	80.2	150	2.9	1965
103. Lansing Smith 2	31	6	62.1	71.8	190	3.0	1967
104. Middletown 3	32	3	67.8	74.8	239	—	1964
105. Mt. Tom 1	33	9	91.0	105.1	136	2.2	1960
106. Baldwin 1	34	3	66.6	71.8	623	3.8	1970
107. Wood River 5	34	9	64.2	69.6	388	3.0	1964
108. Breed 1 †	35	13	59.5	68.1	496	3.7	1960
109. Tanners Creek 4	35	9	58.2	64.1	580	3.3	1964
110. Ratts 1	36	4	65.4	75.5	117	2.8	1969
111. Ratts 2	36	4	64.8	74.9	117	2.6	1969
112. Petersburg 1	37	6	72.1	79.5	253	3.4	1967
113. Petersburg 2	37	4	67.3	77.0	471	3.3	1969
114. Stout 6	37	12	66.8	77.2	114	3.6	1961
115. Kapp 2	38	6	59.8	66.0	218	3.0	1967
116. Prairie Creek B 4	39	5	53.8	62.1	149	2.6	1968
117. Neal 2	40	1	51.3	55.6	349	0.6	1972
118. Burlington 1	41	5	54.1	59.7	212	2.7	1968

Unit	Utility Number	Years In Data Base	Capacity Factor	Capacity Performance	Mw	Avg. Sulfur Content	Year Installed
119. Montrose 2	42	13	66.2	76.5	188	5.9	1960
120. Montrose 3	42	9	65.8	76.0	188	5.8	1964
121. Big Sandy 1	43	11	77.6	85.6	281	0.9	1963
122. Big Sandy 2 †	43	4	59.0	63.5	816	1.0	1969
123. E.W. Brown 2	44	10	50.0	57.7	180	2.7	1963
124. E.W. Brown 3	44	2	50.7	58.0	431	2.1	1971
125. Cane Run 4	45	11	73.6	85.0	163	3.3	1962
126. Cane Run 5	45	7	59.9	66.1	209	3.4	1966
127. Cane Run 6	45	4	47.7	52.7	272	3.5	1969
128. Mill Creek 1	45	1	43.8	47.5	356	3.8	1972
129. Portland 2	46	11	73.5	81.1	255	2.3	1962
130. Milton Young 1	47	3	82.8	91.3	235	0.7	1970
131. Watson 4	48	5	50.1	55.3	299	2.6	1968
132. Sibley 3	49	4	35.9	41.1	419	3.7	1969
133. Fort Martin 1 †	50	6	71.2	78.3	576	3.1	1967
134. Fort Martin 2 †	50	5	72.9	80.2	576	3.0	1968
135. Harrison 1 †	50	1	36.1	38.9	684	4.1	1972
136. Willow Island 2	50	13	84.2	97.2	165	4.2	1960
137. Corette 1	51	5	48.5	56.0	173	0.7	1968
138. Reid Gardner 1	52	8	76.3	88.1	114	0.6	1965
139. Reid Gardner 2	52	5	82.3	95.1	114	0.6	1968
140. Brayton Point 1	53	6	81.7	90.2	241	1.0	1963
141. Brayton Point 2	53	4	89.8	99.1	241	1.0	1964
142. Bailly 7	54	11	66.8	77.2	194	3.8	1962

Unit	Utility Number	Years In Data Base	Capacity Factor	Capacity Performance	Mw	Avg. Sulfur Content	Year Installed
143. Bailly 8 †	54	5	51.4	58.8	422	3.7	1968
144. Mitchell 4	54	3	82.5	95.3	115	3.2	1970
145. High Bridge 6	55	13	84.2	97.2	163	3.2	1960
146. King 1 †	55	5	58.9	64.9	598	3.2	1968
147. Riverside 8	55	9	67.9	74.9	239	2.6	1964
148. Sammis 2	56	13	77.1	89.1	185	2.8	1960
149. Sammis 3	56	12	82.4	95.2	185	2.8	1961
150. Sammis 4	56	11	82.2	95.0	185	2.8	1962
151. Sammis 5	56	6	64.1	69.5	318	2.8	1967
152. Sammis 6 †	56	4	51.9	56.0	623	2.7	1969
153. Sammis 7 †	56	2	54.7	59.0	623	2.8	1971
154. Cardinal 1 †	57	6	69.0	74.4	615	3.2	1967
155. Cardinal 2 †	57	6	69.7	75.2	615	3.2	1967
156. Mitchell 1 †	57	2	47.8	51.5	816	3.5	1971
157. Mitchell 2 †	57	2	66.8	72.0	816	3.5	1971
158. Muskingum River 5†	57	5	66.3	71.4	615	4.9	1968
159. North Omaha 4	58	10	72.1	83.3	136	2.6	1963
160. North Omaha 5	58	5	61.1	67.5	218	2.6	1968
161. Lovett 4	59	2	61.9	71.6	179	1.7	1966
162. Dave Johnston 2	60	12	71.8	83.0	100	0.5	1961
163. Dave Johnston 3	60	9	48.1	53.0	220	0.5	1964
164. Dave Johnston 4	60	1	60.0	65.0	330	0.6	1972
165. Homer City 1 †	61	4	44.4	47.9	660	2.4	1969
166. Homer City 2 †	61	4	45.7	49.3	660	2.4	1969
167. Shawville 4	61	13	79.8	92.2	188	2.9	1960

Unit	Utility Number	Years In Data Base	Capacity Factor	Capacity Performance	Mw	Avg. Sulfur Content	Year Installed
168. Brunner Island 1	62	12	65.7	71.2	363	2.4	1961
169. Brunner Island 2	62	8	69.0	79.0	405	2.4	1965
170. Brunner Island 3 †	62	4	55.7	60.0	790	2.6	1969
171. Conemaugh 1 †*	62	3	54.6	58.9	936	2.6	1970
172. Conemaugh 2 †*	62	2	57.5	62.0	936	2.5	1971
173. Keystone 1 †*	62	6	44.7	48.2	936	2.2	1967
174. Keystone 2†*	62	5	54.0	58.2	936	2.2	1968
175. Montour 1 †	62	1	56.7	61.1	806	1.8	1972
176. New Castle 5	63	9	70.6	81.6	133	3.4	1964
177. Eddystone 1 †	64	13	64.7	70.1	354	2.3	1960
178. Eddystone 2 †	64	13	77.7	84.2	354	2.3	1960
179. Chalk Point 1 †	65	9	56.2	60.9	364	2.2	1964
180. Chalk Point 2 †	65	9	55.3	60.0	364	2.1	1966
181. Dickerson 2	65	13	72.6	83.9	196	2.2	1960
182. Dickerson 3	65	11	77.1	89.1	196	2.2	1962
183. Cherokee 3	66	11	69.0	79.7	171	0.6	1962
184. Cherokee 4	66	5	68.0	73.7	381	0.6	1968
185. Valmont 5	66	9	62.7	72.4	192	0.6	1964
186. Cayuga 1	67	3	54.2	59.7	531	2.3	1970
187. Cayuga 2	67	1	65.0	71.6	531	2.2	1972
188. Gallagher 4	67	12	71.9	83.0	150	3.6	1961
189. Wabash River 6	67	5	55.9	60.6	387	2.6	1968
190. Merrimack 1	68	13	75.4	87.1	114	2.4	1960
191. Merrimack 2	68	5	65.5	71.0	346	2.3	1968
192. Hudson 1 †	69	1	47.9	54.8	455	2.6	1964

*Units 171-174 are owned by Penn P&L but were constructed and are operated by Penn Elec.

Unit	Utility Number	Years In Data Base	Capacity Factor	Capacity Performance	Mw	Avg. Sulfur Content	Year Installed
193. Mercer 1	69	13	68.4	74.1	326	1.9	1960
194. Mercer 2	69	12	67.4	73.0	326	1.9	1961
195. Port Wentworth 3	70	1	63.0	72.8	104	2.5	1965
196. Canadys 1	71	11	66.0	76.3	136	1.2	1962
197. Canadys 2	71	9	64.9	74.9	136	1.2	1964
198. Canadys 3	71	6	38.4	42.4	218	1.2	1967
199. Wateree 1 †	71	3	52.9	57.3	386	1.5	1970
200. Wateree 2 †	71	2	49.4	53.5	386	1.5	1971
201. Jefferies 3	72	4	52.4	60.6	173	1.2	1969
202. Jefferies 4	72	3	65.7	75.9	173	1.0	1970
203. Mohave 1 †	73	2	49.4	53.3	818	0.4	1971
204. Mohave 2 †	73	2	46.0	49.6	818	0.4	1971
205. Gaston 1	74	13	75.4	83.2	272	1.0	1960
206. Gaston 2	74	13	78.2	86.3	272	1.0	1960
207. Gaston 3	74	12	80.6	88.9	272	1.0	1961
208. Gaston 4	74	11	89.1	98.4	245	1.0	1962
209. Big Bend 1	75	3	46.1	52.8	446	3.5	1970
210. Gannon 3	75	13	54.6	63.1	180	3.7	1960
211. Gannon 4	75	9	58.8	67.9	188	3.7	1964
212. Gannon 5	75	8	54.5	60.2	239	3.6	1965
213. Gannon 6	75	6	32.6	37.3	414	3.6	1967
214. Bull Run 1 †	76	6	62.1	66.9	950	1.7	1967
215. Colbert B 1	76	8	55.9	61.5	550	4.1	1965
216. Paradise 1	76	10	67.1	72.3	704	4.1	1963
217. Paradise 2	76	10	67.0	72.2	704	4.1	1963
218. Paradise 3 †	76	3	58.5	63.1	1150	4.1	1970
219. Widows Creek B 1	76	12	59.3	65.2	575	2.8	1961

Unit	Utility Number	Years In Data Base	Capacity Factor	Capacity Performance	Mw	Avg. Sulfur Content	Year Installed
220. Widows Creek B 2	76	8	53.7	59.0	550	3.0	1965
221. Bay Shore 3	77	10	79.5	91.8	141	2.1	1963
222. Bay Shore 4	77	5	68.4	75.5	218	2.1	1968
223. Labadie 1	78	3	44.0	47.4	621	3.1	1970
224. Labadie 2	78	2	46.4	50.0	621	3.1	1971
225. Labadie 3	78	1	58.7	63.3	621	3.1	1972
226. Meramec 4°	78	12	65.3	70.7	359	2.7	1961
227. Sioux 1 †	78	6	44.2	48.6	550	3.0	1967
228. Sioux 2 †	78	5	49.1	54.1	550	3.0	1968
229. Bridgeport Harbor 2	79	5	66.6	76.9	180	—	1961
230. Stanton 1	80	6	68.0	78.5	172	0.8	1967
231. Naughton 1	81	10	65.9	76.1	163	0.5	1963
232. Naugton 2	81	5	69.0	76.1	218	0.5	1968
233. Naughton 3	81	2	49.9	54.1	326	0.6	1971
234. Chesterfield 4	82	9	67.9	78.5	188	1.2	1960
235. Chesterfield 5	82	5	69.7	75.6	359	1.2	1964
236. Possum Point 4	82	6	64.8	71.6	239	0.9	1962
237. Mt. Storm 1	82	8	57.2	62.9	570	2.1	1965
238. Mt. Storm 2	82	7	62.7	69.0	570	2.1	1966
239. Portsmouth 3	82	6	52.8	58.3	239	1.0	1962
240. Portsmouth 4	82	6	63.3	69.9	239	1.0	1962
241. Hatfield's Ferry 1 †	83	4	56.4	62.1	576	2.8	1969
242. Hatfield's Ferry 2 †	83	3	61.1	67.3	576	2.7	1970
243. Hatfield's Ferry 3 †	83	2	63.4	69.7	576	2.5	1971

Unit	Utility Number	Years In Data Base	Capacity Factor	Capacity Performance	Mw	Avg. Sulfur Content	Year Installed
244. Mitchell 3	83	10	74.6	82.4	299	2.8	1963
245. South Oak Creek 6	84	12	55.1	60.9	275	2.3	1961
246. South Oak Creek 7	84	8	61.1	66.2	318	2.2	1965
247. South Oak Creek 8	84	6	65.5	71.0	324	2.2	1967
248. Dewey 2	85	11	67.9	78.4	114	3.2	1962
249. Edgewater 4	85	4	67.0	72.6	351	2.5	1969
250. Pulliam 8	86	9	85.6	98.8	125	2.5	1964

ANNUAL CAPACITY STATISTICS, COAL UNITS 400 Mw OR GREATER

Unit	Mw Capacity	Year	C.F.	C.P.
Barry 4	404	1970	51.3	58.8
"	"	1971	46.8	53.6
"	"	1972	53.9	61.7
"	"	1973	51.6	59.1
Barry 5†	789	1972	33.7	36.3
"	"	1973	52.8	56.9
Gorgas 10†	789	1973	57.9	62.4
Amos 1†	816	1972	80.5	86.8
"	"	1973	79.0	85.1
Amos 2†	816	1973	79.8	86.0
Sporn 5†	496	1962	68.8	78.8
"	"	1963	71.2	81.5
"	"	1964	71.1	81.4
"	"	1965	69.7	79.8
"	"	1966	74.0	84.7
"	"	1967	66.1	75.6
"	"	1968	70.5	80.7
"	"	1969	66.8	76.5
"	"	1970	62.6	71.6
"	"	1971	68.1	77.9
"	"	1972	62.5	71.6
"	"	1973	70.3	80.5
Four Corners 4†	818	1970	51.1	55.1
"	"	1971	64.1	69.1
"	"	1972	70.7	76.2
"	"	1973	65.0	70.1
Four Corners 5†	818	1971	55.2	59.5
"	"	1972	66.3	71.5
"	"	1973	68.5	73.8
New Madrid 1	650	1973	49.1	52.9
Roxboro 1	411	1967	76.1	87.1
"	"	1968	71.1	81.4
"	"	1969	66.9	76.5
"	"	1970	64.9	74.3
"	"	1971	65.6	75.1
"	"	1972	62.0	71.0
"	"	1973	66.8	76.5
Roxboro 2	657	1969	65.9	71.0
"	"	1970	70.2	75.7
"	"	1971	73.8	79.6
"	"	1972	72.6	78.2
"	"	1973	70.1	75.6

Unit	Mw Capacity	Year	C.F.	C.P.
Coffeen 2	617	1973	42.3	45.5
Beckjord 6	461	1970	52.9	60.6
"	"	1971	62.1	71.1
"	"	1972	55.5	63.5
"	"	1973	33.9	38.8
Avon 9†	680	1971	71.7	77.3
"	"	1972	63.9	68.9
"	"	1973	59.4	64.1
Eastlake 5†	680	1973	66.2	71.3
Stuart 2†	610	1972	57.7	62.2
"	"	1973	66.3	71.5
Stuart 1†	610	1971	55.2	59.5
"	"	1972	67.0	72.2
Stuart 3†	610	1973	76.2	82.1
Joliet 7	660	1966	53.7	57.8
"	"	1967	58.4	62.9
"	"	1968	46.3	49.9
"	"	1969	59.2	63.8
"	"	1970	48.8	52.6
"	"	1971	43.8	47.2
"	"	1972	50.4	54.3
"	"	1973	53.8	58.0
Joliet 8	660	1967	55.0	59.3
"	"	1968	63.6	68.6
"	"	1969	44.8	48.3
"	"	1970	54.0	58.2
"	"	1971	49.7	53.6
"	"	1972	50.7	54.6
"	"	1973	54.2	58.4
Kinkaid 1	660	1968	34.1	36.8
"	"	1969	40.6	43.8
"	"	1970	48.7	52.5
"	"	1971	40.1	43.2
"	"	1972	48.3	52.1
"	"	1973	41.6	44.9
Kinkaid 2	660	1969	41.9	45.1
"	"	1970	49.6	53.4
"	"	1971	52.6	56.7
"	"	1972	46.0	49.6
"	"	1973	42.5	45.8
Powerton 5	893	1973	51.6	55.6
Will County 4	598	1964	62.4	68.7
"	"	1965	51.9	57.1
"	"	1966	49.0	53.9

Unit	Mw Capacity	Year	C.F.	C.P.
"	"	1967	49.9	54.9
"	"	1968	59.0	64.9
"	"	1969	56.3	62.0
"	"	1970	53.3	58.6
"	"	1971	52.9	58.2
"	"	1972	36.7	40.4
"	"	1973	49.9	54.9
Ravenswood 3	1028	1967	41.4	44.6
"	"	1968	41.0	44.2
"	"	1969	46.6	50.3
Big Brown 1†	593	1973	65.7	72.3
Big Brown 2†	593	1973	63.0	69.4
Monroe 1†	817	1972	49.1	52.9
"	"	1973	55.6	59.9
St. Clair 7	545	1970	74.3	81.8
"	"	1971	71.4	78.5
"	"	1972	69.8	76.8
"	"	1973	64.9	71.5
Trenton Channel 9	536	1969	79.6	87.6
"	"	1970	67.4	74.2
"	"	1971	80.0	88.1
"	"	1972	81.8	90.0
"	"	1973	72.9	80.2
Cliffside 5	571	1973	69.4	76.4
Marshall 3†	650	1970	60.4	65.2
"	"	1971	74.0	79.7
"	"	1972	77.0	83.0
"	"	1973	82.3	88.8
Marshall 4†	650	1971	70.9	76.4
"	"	1972	72.1	77.8
"	"	1973	87.4	94.2
Cheswick 1	565	1971	59.4	65.4
"	"	1972	65.3	71.9
"	"	1973	67.6	74.4
Crystal River 1	440	1967	54.2	62.0
"	"	1968	53.8	61.6
"	"	1969	52.6	60.2
Crystal River 2	524	1970	48.8	53.7
"	"	1971	40.5	44.5
Bowen 1†	806	1972	43.4	46.7
"	"	1973	43.9	47.3
Bowen 2†	806	1973	53.1	57.3
Hammond 4	578	1971	37.0	40.8
"	"	1972	38.5	42.4
"	"	1973	43.5	47.8

Unit	Mw Capacity	Year	C.F.	C.P.
Harllee Branch 3†	544	1969	53.2	58.5
"	"	1970	58.9	64.8
"	"	1971	52.3	57.5
"	"	1972	57.5	63.3
"	"	1973	38.0	41.9
Harllee Branch 4†	544	1970	57.6	63.4
"	"	1971	47.6	52.4
"	"	1972	34.4	37.8
"	"	1973	32.3	35.6
Baldwin 1	623	1971	62.2	67.0
"	"	1972	69.4	74.8
"	"	1973	68.3	73.6
Breed 1†	496	1961	68.0	77.8
"	"	1962	68.8	78.7
"	"	1963	72.1	82.6
"	"	1964	62.0	71.0
"	"	1965	72.3	82.8
"	"	1966	72.6	83.2
"	"	1967	62.8	71.9
"	"	1968	57.1	65.4
"	"	1969	61.6	70.5
"	"	1970	51.9	59.5
"	"	1971	45.7	52.3
"	"	1972	15.3	17.5
"	"	1973	62.6	71.7
Tanners Creek 4†	580	1965	67.7	74.5
"	"	1966	81.2	89.4
"	"	1967	55.7	61.3
"	"	1968	48.3	53.2
"	"	1969	77.7	85.5
"	"	1970	47.0	51.7
"	"	1971	29.5	32.5
"	"	1972	52.3	57.6
"	"	1973	64.3	70.8
Petersburg 2	471	1970	66.8	76.5
"	"	1971	71.7	82.1
"	"	1972	67.0	76.7
"	"	1973	63.4	72.6
Big Sandy 2†	816	1970	50.8	54.8
"	"	1971	65.4	70.5
"	"	1972	69.6	75.0
"	"	1973	50.0	53.8
E.W. Brown 3	431	1972	38.4	44.0

Unit	Mw Capacity	Year	C.F.	C.P.
"	"	1973	63.0	72.1
Sibley 3	419	1970	34.7	39.7
"	"	1971	27.8	31.8
"	"	1972	44.1	50.5
"	"	1973	37.1	42.5
Fort Martin 1†	576	1968	66.6	73.3
"	"	1969	78.6	86.5
"	"	1970	76.5	84.2
"	"	1971	60.2	66.3
"	"	1972	75.7	83.3
"	"	1973	69.4	76.3
Fort Martin 2†	576	1969	77.8	85.6
"	"	1970	70.4	77.4
"	"	1971	73.5	80.9
"	"	1972	70.8	77.9
"	"	1973	72.1	79.4
Harrison 1†	684	1973	36.1	38.9
Bailly 8†	422	1969	57.9	66.3
"	"	1970	52.8	60.5
"	"	1971	44.7	51.2
"	"	1972	47.2	54.0
"	"	1973	54.2	62.0
King 1†	598	1969	54.4	59.9
"	"	1970	62.2	68.5
"	"	1971	57.9	63.7
"	"	1972	63.0	69.3
"	"	1973	57.1	62.9
Sammis 6†	623	1970	51.0	55.0
"	"	1971	24.6	26.6
"	"	1972	60.6	65.4
"	"	1973	71.4	76.9
Sammis 7†	623	1972	59.2	63.8
"	"	1973	50.2	54.1
Cardinal 1†	615	1968	79.9	86.1
"	"	1969	71.0	76.5
"	"	1970	80.5	86.8
"	"	1971	57.8	62.3
"	"	1972	59.2	63.8
"	"	1973	65.8	70.9
Cardinal 2†	615	1968	79.0	85.2
"	"	1969	77.1	83.1
"	"	1970	61.0	65.7
"	"	1971	57.2	61.7
"	"	1972	73.7	79.4

Unit	Mw Capacity	Year	C.F.	C.P.
"	"	1973	70.4	75.9
Mitchell 1†	816	1972	29.4	31.7
"	"	1973	66.2	71.4
Mitchell 2†	816	1972	66.7	71.9
"	"	1973	66.8	72.0
Muskingum River 5†	615	1969	60.4	65.1
"	"	1970	62.9	67.8
"	"	1971	72.1	77.7
"	"	1972	68.4	73.7
"	"	1973	67.6	72.9
Homer City 1†	660	1970	40.7	43.8
"	"	1971	29.9	32.3
"	"	1972	49.5	53.4
"	"	1973	57.5	61.9
Homer City 2†	660	1970	43.1	46.5
"	"	1971	43.5	46.9
"	"	1972	50.1	54.0
"	"	1973	46.1	49.7
Brunner Island 2	405	1966	63.4	72.6
"	"	1967	69.5	79.6
"	"	1968	79.0	90.5
"	"	1969	70.6	80.8
"	"	1970	69.1	79.1
"	"	1971	63.4	72.6
"	"	1972	61.9	70.8
"	"	1973	74.9	85.7
Brunner Island 3†	790	1970	35.1	37.8
"	"	1971	61.1	65.8
"	"	1972	59.6	64.2
"	"	1973	67.1	72.3
Conemaugh 1†	936	1971	51.4	55.4
"	"	1972	54.1	58.3
"	"	1973	58.4	62.9
Conemaugh 2†	936	1972	55.0	59.3
"	"	1973	60.1	64.7
Keystone 1†	936	1968	44.2	47.6
"	"	1969	41.2	44.5
"	"	1970	30.1	32.4
"	"	1971	41.3	44.5
"	"	1972	51.9	56.0
"	"	1973	59.6	64.3
Keystone 2†	936	1969	61.3	66.0
"	"	1970	41.8	45.0
"	"	1971	58.1	62.6

Unit	Mw Capacity	Year	C.F.	C.P.
"	"	1972	49.5	53.3
"	"	1973	59.4	64.0
Montour 1†	806	1973	56.7	61.1
Cayuga 1	531	1971	46.7	51.4
"	"	1972	56.3	61.9
"	"	1973	59.7	65.7
Cayuga 2	531	1973	65.0	71.6
Hudson 1†	455	1968	47.9	54.8
Mohave 1†	818	1972	45.1	48.6
"	"	1973	53.7	57.9
Mohave 2†	818	1972	47.6	51.3
"	"	1973	44.4	47.9
Big Bend 1	446	1971	37.3	42.8
"	"	1972	50.5	57.8
"	"	1973	50.5	57.8
Gannon 6	414	1968	24.9	28.5
"	"	1969	15.7	18.0
"	"	1970	36.4	41.7
"	"	1971	37.0	42.4
"	"	1972	44.0	50.4
"	"	1973	37.6	43.1
Bull Run 1†	950	1968	64.6	69.6
"	"	1969	49.7	53.6
"	"	1970	65.7	70.8
"	"	1971	60.6	65.3
"	"	1972	60.9	65.5
"	"	1973	70.8	76.3
Colbert B 1	550	1966	61.7	68.9
'	"	1967	49.8	54.8
"	"	1968	48.0	52.9
"	"	1969	61.7	67.9
"	"	1970	49.8	54.8
"	"	1971	48.2	53.0
"	"	1972	67.8	74.6
"	"	1973	60.2	66.2
Paradise 1	704	1964	44.3	47.7
"	"	1965	68.9	74.2
"	"	1966	69.7	75.1
"	"	1967	76.9	82.9
"	"	1968	78.0	84.1
"	"	1969	72.3	77.9
"	"	1970	72.9	78.5
"	"	1971	64.1	69.1
"	"	1972	62.9	67.8

Unit	Mw Capacity	Year	C.F.	C.P.
"	"	1973	60.5	65.2
Paradise 2	704	1964	58.7	63.2
"	"	1965	57.2	61.6
"	"	1966	70.2	75.6
"	"	1967	69.5	74.9
"	"	1968	76.7	82.6
"	"	1969	67.1	72.4
"	"	1970	67.6	72.9
"	"	1971	70.2	75.6
"	"	1972	66.9	72.1
"	"	1973	65.6	70.7
Paradise 3†	1150	1971	50.7	54.6
"	"	1972	62.5	67.4
"	"	1973	62.3	67.2
Widows Creek B 1	575	1962	62.8	69.1
"	"	1963	80.2	88.2
"	"	1964	76.5	84.2
"	"	1965	50.2	55.3
"	"	1966	65.1	71.6
"	"	1967	73.8	81.2
"	"	1968	52.2	57.4
"	"	1969	61.6	67.8
"	"	1970	51.4	56.5
"	"	1971	49.1	54.0
"	"	1972	44.3	48.7
"	"	1973	44.2	48.6
Widows Creek B 2	550	1966	53.2	58.6
"	"	1967	56.8	62.5
"	"	1968	54.7	60.2
"	"	1969	46.6	51.3
"	"	1970	56.9	62.7
"	"	1971	65.1	71.6
"	"	1972	56.5	62.2
"	"	1973	39.2	43.2
Labadie 1	621	1971	55.6	59.9
"	"	1972	13.6	14.6
"	"	1973	62.8	67.7
Labadie 2	621	1972	44.9	48.4
"	"	1973	47.9	51.7
Labadie 3	621	1973	58.7	63.3
Sioux 1†	550	1968	46.3	50.9
"	"	1969	47.4	52.1
"	"	1970	59.7	65.6
"	"	1971	49.6	54.6

Unit	Mw Capacity	Year	C.F.	C.P.
"	"	1972	35.2	38.7
"	"	1973	26.9	29.6
Sioux 2†	550	1969	54.7	60.2
"	"	1970	57.8	63.7
"	"	1971	48.7	53.6
"	"	1972	41.6	45.8
"	"	1973	42.8	47.1
Mt. Storm 1	570	1966	39.6	43.6
"	"	1967	62.0	68.2
"	"	1968	56.3	62.0
"	"	1969	69.5	76.4
"	"	1970	58.9	64.8
"	"	1971	72.2	79.4
"	"	1972	56.7	62.4
"	"	1973	42.5	46.7
Mt. Storm 2	507	1967	45.4	49.9
"	"	1968	70.8	77.9
"	"	1969	62.3	68.5
"	"	1970	63.6	70.0
"	"	1971	68.3	75.2
"	"	1972	68.4	75.3
"	"	1973	60.3	66.4
Hatfield's Ferry 1†	576	1970	61.0	67.2
"	"	1971	49.8	54.8
"	"	1972	55.3	60.9
"	"	1973	59.4	65.4
Hatfield's Ferry 2†	576	1971	62.2	68.4
"	"	1972	55.9	61.5
"	"	1973	65.3	71.9
Hatfield's Ferry 3†	576	1972	63.8	70.3
"	"	1973	62.9	69.2

Appendix B: The Definition of Capacity Factor

Capacity factor is defined in the power industry as the ratio of (i) a unit's electrical generation during a period to (ii) the product of the unit's rated generating capacity and the number of hours in the period. Or,

$$\text{Capacity Factor} = \frac{\text{generation (in Mw-hours)}}{\text{capacity (in Mw)} \times \text{hours in period}}$$

Over the course of a year, capacity factor is computed by:

$$\text{Capacity Factor} = \frac{\text{generation}}{\text{capacity} \times 8760},$$

except in leap years, for which 8760 is replaced by 8784 (the number of hours in a leap year).

Capacity factor is thus a ratio, expressed as a percentage, of a generating unit's electrical generation to its maximum possible generation during a period.

The terms *generation* and *capacity* can each be defined in several ways. This variation has given rise to several different definitions of capacity factor, and competing sets of statistics as well, for both nuclear and fossil units.

For both nuclear and fossil units, we have defined generation in this study as *net generation*. This is the amount of electricity produced by the power plant after subtracting the electricity consumed by the plant itself, to operate pumps, heaters, motors, control equipment, and so forth. Or, net generation is the amount of electricity delivered to the "bus-bar" by the plant and made available for transmission into the utility grid system. It is a well-defined, unambiguous term reported annually by the utilities to both the NRC and FPC, and is used in virtually all capacity factor tabulations.* Net generation is also the basis of all computations of fuel and operating costs, such as "heat rate" (thermal efficiency), and per-kwh fuel and O&M expenses.

*The only exception of which we are aware is the data base of the Equipment Availability Task Force of the Edison Electric Institute, which computes capacity factor from gross generation and a measure of capacity known as "gross maximum dependable capability."

For nuclear units, we have measured capacity—the other term which enters into the computation of capacity factor—as the *design electrical rating* reported by utilities to the NRC and published in the monthly NRC Gray Books. Design electrical rating is defined by the NRC as the "nominal net electrical output of the unit specified by the utility and used for the purpose of plant design." The design rating is thus defined analogously to net generation, in that both exclude power generation which is required for normal in-plant use. Use of the two measurements together thus represents consistent practice.*

For coal-fired units, we have measured capacity as the *installed generating capacity* reported by utilities to the FPC in the annual Form 1 report (page 432, line five; installed generating capacity also appears on line one of the annual FPC Steam-Electric Plant Cost Books).

The installed generating capacity is nominally defined by the FPC to be the "Maximum Generator Nameplate Rating." This is the peak generator output under *normal* operating conditions. Higher outputs can sometimes be produced for brief periods of up to several hours by operating the generator at higher coolant pressure. Maximum generator nameplate rating generally includes some or all of the unit's power capability which is consumed to operate plant equipment such as pumps, fans, and heaters. Therefore, it is closer to a gross rating rather than a net rating, such as the design electrical rating which we use to measure nuclear unit capacity.

We have selected these measures of capacity—NRC design electrical rating for nuclear units, FPC installed generating capacity for coal units—because these are the capacity ratings most commonly used in the power industry to compute per-Kw capital costs. (As we discuss in Chapter 9, power plant per-Kwh generating costs are computed by allocating annual capital costs among the Kwh's generated, which is measured by capacity factor. The same capacity measure used to compute per-Kw capital costs should be used to compute capacity factor. This point is discussed further below.) For example, in the FPC Steam-Electric Plant Cost Books, the "Cost Per Kilowatt of Installed Capacity"—the \$/Kw shorthand measure of capital cost—is explicitly computed by dividing total plant cost by installed generating capacity.**

While the FPC reports the installed generating capacity of nuclear as well as fossil units, we have used the NRC design electrical rating rather than the FPC rating to measure nuclear capacity. This is because the NRC design rating is the capacity most frequently used in computing per-Kw costs for nuclear units. Since the average FPC capacity rating of nuclear units in the CEP data base is

*The design ratings reported by utilities for several nuclear units in the CEP data base are apparently gross output ratings, in that they include generating capacity which is dedicated to in-plant use. Nevertheless, we have used these ratings in order to maintain a consistent non-discretionary procedure for compiling nuclear capacity figures, namely utility reports in the Gray Books, and to use capacity figures which are consistent with economic analysis. This latter point is discussed below in the text.

**The same point is made in a recent paper by George Rathjens and D. Schiff of MIT: "Since the capital charges for generating electricity are proportional to cost per Kw for installed plant divided by the capacity factor, it would obviously make no difference which definition of plant rating were used in estimating the cost of power generation, provided the same definition was used in calculating plant cost. Since this appears usually to be based on design rating, that would seem to be the logical figure to use in calculating capacity factors." "The Economics of Nuclear & Coal Generated Power." Rathjens & Schiff, July 13, 1976.

about 1.5% *higher* than the average NRC rating, use of the NRC rating in this study results in an average nuclear capacity factor which is 1.5% higher, or about 0.9 percentage points higher, than the capacity factor using the FPC rating.*

Unfortunately, there is some variability in the way utilities define the installed capacity of their coal-fired units. While the variability is probably ''washed out'' in computing aggregate average capacity factors for coal units, it may introduce errors in comparing units of different size, coal quality, ownership, etc. within the coal unit sample.

For example, whereas most utilities use maximum generator nameplate rating to represent installed generating capacity, some use other measures, chiefly maximum turbine ratings. Minnkota Power Corp. reports the installed generating capacity of its Milton R. Young No. 1 unit as 234.6 Mw, which is the maximum turbine rating, rather than reporting the maximum generator rating of 256.5 Mw.

Conversely, the generator rating may be higher than the turbine rating yet still be reported by the utility as the installed capacity, e.g., Carolina P&L's 207-Mw Asheville 2 station, which has a maximum turbine rating of 186 Mw. Regardless of the relationship between turbine and generator ratings (typically they are rated equally), we have used the installed generating capacity figure as the coal-fired unit's capacity.**

Recently, several organizations have begun computing capacity factors for nuclear units on the basis of *maximum dependable capability* (MDC), in addition to the NRC design rating. As its name implies, MDC is the maximum output that can be relied upon from a generating plant. MDC was originally formulated to measure the modest reductions in plant capability which may occur in summer months. (Turbine-generator efficiency is affected by the temperature of cooling water in the condenser, with higher temperatures resulting in reduced efficiency.) Since MDC is always equal to or less than the design electrical rating, capacity factors computed using MDC are always equal to or greater than capacity factors computed from the design rating.

*Other observers have defined nuclear and coal capacities in ways that are more favorable to coal than our definitions in a strict capacity factor comparison. For example, Nuclear Services Corporation, in its power plant reliability study for FEA, lists capacities for a sample of five large nuclear units, and nine large coal-fired units in which the average nuclear capacity is 1% *higher* than the average capacity used by CEP, while the average coal unit capacity is 8% *less* than the average capacity as measured by CEP. The result of our definition of capacity is then to add 9% to the average nuclear capacity factor relative to coal, for the particular sample. (Nuclear Services Corporation, ''Productivity Improvement Action Plan,'' November 28, 1975, p. 16)

**There is sufficient variability among the NRC design ratings reported for nuclear units that one utility engineer, Gene Cramer of Southern California Edison, has proposed using the *thermal* rather than the *electrical* power output of nuclear units to measure capacity factor. Cramer notes that the nuclear licensing process places a limit upon the reactor's thermal power output, allowing the plant to make whatever electrical power is possible. The thermal output is therefore a fairly straightforward physical quantity, while the electrical rating is assigned according to the utility's discretion. Cramer cites several instances in which similarly designed reactors with similar thermal output are reported with different design electrical ratings. (Private communication, July 22, 1976.)

Unfortunately, reactor thermal output suffers from certain deficiencies as a basis for computing capacity factor, such as the absence of a standard yardstick for converting thermal output into electrical. The use of the NRC design electrical rating to measure capacity factor may introduce small errors in capacity factor comparisons among nuclear units, but it appears to result in an accurate and meaningful measure of capacity factor for nuclear units as a whole.

Several utilities currently estimate MDC for nuclear units on the basis of mechanical or regulatory conditions which restrict output. For example, Northeast Utilities' Millstone 2 nuclear unit, with a net design rating of 828 Mw, has been assigned an MDC of 780 Mw "based on the unavailability of the moisture separator/reheaters."* In a more extreme example, Commonwealth Edison has assigned an MDC of 225 Mw to its 809-Mw Quad Cities 1 and 2 nuclear units, on the basis of reduced summer capabilities caused by problems with the spray canal condenser cooling system required to minimize thermal impact on the Mississippi River. This has resulted in Commonwealth Edison's reporting capacity factors up to 300% for Quad Cities during the winter of 1975-76, since generation from units capable of operating at or close to 809 Mw is being compared against a "capability" of only 225 Mw.**

While there is a legitimate use for MDC by organizations such as the regional Electric Reliability Councils, which are responsible for monitoring power plant availability to maintain reliable electric systems, we believe that capacity factor measurements based on the design electrical capacity (and net generation) are the most legitimate measurements for economic evaluation of nuclear units. As a rule, utilities perform power plant economic evaluations using estimates of net generation derived by applying a projected capacity factor to the *design electrical rating* (in the case of a nuclear unit). Application of a capacity factor projected on the basis of capacity factors computed from many MDC ratings, to a design electrical capacity, is generally inconsistent and would result in an overestimate of plant generation.

Similarly, economic evaluation involves estimating capital costs, and dividing estimated capital cost by projected net generation to estimate capital charges per unit of generation. Capital costs are commonly estimated in terms of $/Kw, where Kw is the net design rating. Net generation projected using capacity factors computed from many MDC ratings would be inconsistent with $/Kw computed with net design rating and would lead to an underestimate of per-unit capital charges.

Finally, the net design rating is superior to MDC since the former is specified when the plant is designed and remains constant throughout the life of the plant, regardless of the actual level of operating limitations required of the plant. This is consistent with the fact that the fixed charges on a plant remain constant once the plant has been constructed. MDC, on the other hand, may fluctuate as shown in the examples cited earlier.***

The Edison Electric Institute defines capacity factor as the ratio of (i) gross generation to (ii) the product of gross maximum dependable capability and the number of hours. The use of gross measures for both generation and capacity is internally consistent, and in fact the EEI measure of capacity factor may be useful for analyzing the mechanical reliability of a generating unit. However, our definition of capacity factor is more appropriate for economic analysis, since it is consistent with measurements used in estimating plant costs.

*Source: April 1976 Gray Book, NRC, p. 2-57.

**Source: January 1976 Gray Book, NRC, p. 2-85, and subsequent Gray Books.

***The CEP nuclear data base listing in Appendix A gives the NRC design rating, FPC nameplate rating, and MDC of each nuclear unit. For coal-fired units, only the FPC nameplate rating is listed since the design rating is not defined for coal units and MDC ratings are not readily available.

Appendix C: Definition and Measurement of Capacity Performance

In this Appendix we describe the model we have developed to estimate the operating reliability or *capacity performance* of coal-fired units, based on their capacity factors and the extent of load-following. We also present the estimates of load-following which we have used to develop the measures of capacity performance which are used for coal-fired units in this study.

1. Capacity Factor and Load-Following

Nuclear plants have been consistently base-loaded—operated at their full available capacities—by U.S. utilities to date. This is a result of two factors. First, the incremental operating costs of nuclear units have been significantly less than those of other units, except for hydro-electric units.* Second, utilities have rarely experienced periods in which the capacity required to serve customer load has been less than the sum total of available nuclear capacity plus the minimum load desirable on fossil units for maintenance and durability considerations.**

*Hydroelectric stations currently provide about 15% of U.S. electrical generation, not all of which has zero incremental cost since many hydro stations are supplied by reservoirs with limited storage capability. Moreover, more than one-third of U.S. hydro generation is produced in the Pacific Northwest, whose first nuclear unit (Trojan) did not enter commercial service until December 1975 and thus is not included in the CEP data base. Thus, cheaper hydro power has not led to significant load-following by nuclear units in the data base.

**Commonwealth Edison, with seven nuclear units totalling 5536 Mw capacity in commercial service by the end of 1974, is sometimes cited as a utility with sufficient nuclear capacity (along with substantial modern fossil capacity which should be kept above a specified minimum load) that its nuclear units are sometimes deliberately operated at less than full capacity. However, Comm Ed's nuclear units have been out of service with such frequency that the extent of nuclear load-following has been much less than might be anticipated. Comm Ed has estimated that intentional load reduction was imposed on its nuclear units only "four to five times" in 1974, and lasted no more than 15 hours on any occasion. Nuclear load-following had not occurred in off-peak seasons, since many units had been removed from service for maintenance, but had occurred several times after long heat waves, when all equipment was available and in service. (Telephone communication with Nate Kershaw, Comm Ed Administrative Engineer, June 20, 1975.)

As we noted in Section 4.5, the average load-following by nuclear plants has been equivalent to no more than 1% of capacity factor. As a result, nuclear plant capacity factors are an extremely accurate measure of the operating reliability of nuclear plants to date.

Coal-fired units, however, do tend to "follow load" to some extent. That is, the level of operation of coal-fired units has been influenced by the system demand for electricity as well as by the mechanical availability of the units themselves. The actual amount of load-following varies of course from unit to unit and from season to season, as a function of (i) the unit's coal costs and generating efficiency relative to other units; (ii) the unit's operating flexibility in being brought from shut-down to start-up or from low to high load; (iii) the system electric load; (iv) the geographic distribution of customer demand and transmission capability; and (v) other considerations such as incremental maintenance costs and relative environmental impact. The effect of load-following is to cause a unit's capacity factor to be less than the unit's operating reliability, since load-following represents generation which the unit presumably could have produced had the utility required it.

Since we are concerned in this study with estimating power plant operating reliability as a means of projecting the capacity factors of new generating units, it is necessary to develop a measure of operating reliability which takes into account load-following as well as capacity factor.

2. The Definition of Capacity Performance

Given that a power plant was not fully base-loaded during a period (*i.e.*, that it followed load to some extent), we wish to estimate what its operating reliability was during the period. Or, we wish to measure the percentage of the capacity factor sought from the unit by the operating utility which was actually achieved by the unit. We can define this percentage as "capacity performance," as follows:

(Equation C1)

$$\text{Capacity Performance} = \frac{\text{Capacity Factor}}{\text{Optimal Capacity Factor}}$$

where optimal capacity factor is defined to be the capacity factor which the unit would have achieved had it always been available to operate at the level sought by the utility each hour of the year, *i.e.*, had it not been subject to any mechanical outages. Optimal capacity factor is then equal to 100% minus the capacity factor "foregone" by a generating unit because "system conditions" (low system load, or perhaps transmission line considerations) dictated that the unit should not be fully base-loaded.

We noted above that nuclear units to date have "foregone" less than 1% of their potential capacity factor, on the average, because of load following, so that their optimal capacity factor has averaged 99% or higher. The average optimally reliable nuclear unit would thus have a 99% capacity factor. Coal units typically forego *more* of their potential generation to load-following, so that their average optimal capacity factor is *less* than 99%, as we quantify below.*

Referring to Equation C1, we see that capacity performance measures a generating unit's electrical output as a percentage of the output sought from the unit by the operating utility. Capacity performance therefore approximates the capacity factor which the unit would have achieved, had it been fully base-loaded. The capacity factor of existing generating facilities can then be used as a basis for projecting the capacity factor of future base-loaded facilities.

3. An Example of Capacity Performance

As we discuss in Section 5 of this Appendix, a typical coal-fired unit in the CEP data base has had an optimal capacity factor of roughly 90%. That is, a typical coal-fired unit would have been unneeded to generate an amount of electricity equivalent to a 10% capacity factor, because of intentional reductions in output dictated by the operating utility.

It is likely that not all of the average 10% capacity factor foregone by a coal-fired unit would have been realized as generation had the unit been fully base-loaded. Some of the 10% might have been experienced as full or partial forced outages, in the same way that the unit was subject to such outages during the hours in which it *was* base-loaded. If a unit's capacity factor was 60% and its optimal capacity factor 90%, so that the unit achieved ⅔ of the generation sought by the utility, it can be assumed as a first approximation that, had the unit been fully base-loaded, it would have produced ⅔ of the 10% worth of generation for which it was unavailable because of load-following. Its capacity factor, under full base-loading, would have been 60%, plus ⅔ × 10%, or 60% + 6.7%, which is 66.7%.

The same result is obtained from Equation C1 defining capacity performance as the ratio of capacity factor to optimal capacity factor, or 60%/90%, or 66.7%. This is the basis of the definition of capacity performance and our assertion that capacity performance is an estimate of the capacity factor a unit would have achieved had it been base-loaded, and is therefore the appropriate measure of power plant reliability.**

From our conclusion that average load-following by *nuclear* units in the CEP data base has been equivalent to no more than 1% of a year's capacity factor (so that optimal capacity factor for nuclear units has averaged 99% or higher), it is seen that the capacity performance of nuclear units is typically only

*Optimal capacity factor should not be confused with estimates of "attainable" capacity factor. A nuclear unit for which 99% is the optimal capacity factor level (since its full generation is desired by the utility all but 1% of the time) may be limited to an annual attainable capacity factor of 88% if it must be shut down for refueling six weeks of every year. During those six weeks, the utility would still wish to operate the nuclear unit, although it is of course unavailable. Attainable capacity factor is related to mechanical reliability, while optimal capacity factor is an economic concept.

**Continuing this hypothetical example, it is likely that the unit would have generated additional electricity equivalent to *more* than ⅔ of 10% capacity factor, had it been fully base-loaded. This is because the ⅔ ratio of capacity factor to optimal capacity factor reflects generation lost for scheduled maintenance as well as forced outages, whereas load-following explicitly implies no scheduled maintenance. However, the formulation of capacity performance in the text assumes that some of the generation foregone for load-following would in effect be lost for scheduled maintenance. It is therefore a conservative formulation, in that it results in lower capacity performance estimates.

marginally higher than capacity factor. Since the average capacity factor of nuclear units in the CEP data base is 59.3%, the average capacity performance of nuclear units would be no higher than 59.9% (59.3%/99%) or at most only 0.6 percentage points higher than capacity factor. Accordingly, we have not applied capacity performance measurements to nuclear units.

The difference in optimal capacity factor between nuclear and coal-fired units merely reflects the fact that nuclear running costs are so much lower than other units' that it has rarely been economically desirable to operate nuclear units at less than full capacity. In the future, as more nuclear units are installed, or if utilities install other units with operating costs less than those of today's reactors (e.g., breeder reactors, high-temperature gas-cooled reactors, wind generators), or if nuclear fuel costs increase significantly, there will be periods in which conventional reactors will be intentionally reduced in load, as coal-fired units are at various times of the year.

4. A Method for Estimating Capacity Factor

Unfortunately, utilities do not measure optimal capacity factor, which after all is a theoretical construct. Therefore, capacity performance cannot be determined precisely for individual units. However, the Edison Electric Institute does maintain statistics from which it is possible to estimate the average optimal capacity factor of classes of generating units, beginning with the year 1968. The method for developing such estimates is described below.

Equation C1 defining capacity performance may be rewritten as follows:

$$CP = \frac{CF}{OCF}, \text{ or}$$

(Equation C2)

$$CP = \frac{CF}{1 - EEOH/PH}$$

CP is Capacity Performance
CF is Capacity Factor
OCF is Optimal Capacity Factor
EEOH is "Equivalent Economy Outage Hours"
PH is "Period Hours" (*i.e.*, hours in the year or period in question)

EEOH, or equivalent economy outage hours, is the generation lost to load-following, so that EEOH/PH is the capacity factor foregone because of load-following. (In the example in Section C, EEOH/PH equalled 10%)

Load-following can occur in either of two ways: by reducing the output of a unit (which we refer to as "partial economy outage hours") or by shutting the unit down entirely because of insufficient electric demand ("full economy outage hours"). Partial economy outage hours represent hours in which a unit lost some, but not all, of its potential generation due to load-following. Such hours must be "equivalized"—converted to an estimate of equivalent full economy outage hours— to enable us to estimate optimal capacity factor.

For example, a unit may have been on economy shut-down for 276 hours during a year. During those hours the unit was fully available, but was not operated, presumably because of the ability of other generating units with lower incremental operating costs to supply the necessary system load.

In addition, during 1500 other hours the hypothetical unit may have been operated, but at an average load of only 60% of rated capacity due to fluctuations in the utility's need for generation from the unit. These "partial economy outages," averaging 40% of capacity, would have an "equivalized value" of 600 hours (1500 hours × 40% de-rating).

Then the total equivalent economy outage hours for the year for this unit would be 876: 276 full economy outage hours plus 600 equivalized partial economy outage hours. The unit's optimal capacity factor for the year was then:

$$100\% - \frac{876}{8760}, \text{ or } 90\%.$$

This 90% figure measures the capacity factor which the unit would have achieved had it been able to generate the amount of electricity sought by the utility each hour of the year.

Referring back to Equation C2, Equivalent Economy Outage Hours, or EEOH, can be rewritten as

(Equation C3)

$$\text{EEOH} = \text{FEOH} + \text{EPEOH}$$

Equivalent Economy Outage Hours = Full Economy Outage Hours + Equivalized Partial Economy Outage Hours.

FEOH is reported for classes of generating units by the Edison Electrical Institute in its annual "Reports on Equipment Availability." It is generally computed by EEI as the difference between "available hours" and "service hours" for a generating unit.

EPEOH may be approximated as follows:

(Equation C4)

$$\text{EPE OH} = (\text{EA} - \text{CF}) \times \text{PH}$$

EA = "Equivalent Availability"
CF = Capacity Factor
PH = Period Hours

Since these quantities are also reported by EEI in its data summaries, it is possible to use the EEI data to obtain average values of EPEOH for different classes of units.

Equation C4 was derived as follows:

EEI defines and computes Equivalent Availability:

$$\text{EA} = \frac{\text{Service Hours} - \text{Equivalized Outage Hours,}}{\text{PH}}$$

where equivalized outage hours are the hours in which the unit is operated only partially, due to mechanical outages (planned and forced), with the hours "equivalized" according to the percent de-rating during each partial outage. Equivalized outage hours are thus equal to the number of hours in which the unit was operating only partially due to mechanical outages, minus the "equivalized operation" (mwhrs/mw) during those hours.

Or, Equivalized Outage Hours = B − EB

B = Number of hours of partial operation due to mechanical outages,

EB = Equivalized operation (in hours) during those hours.

$$\text{Therefore, EA (Equivalent Availability)} = \frac{\text{SH} - \text{Equivalized Outage Hours}}{\text{PH}},$$

$$\text{or} \quad \text{EA} = \frac{\text{SH} - \text{B} + \text{EB}}{\text{PH}}.$$

Now, service hours can be regarded as the sum of (1) the hours in which the unit operated at full capacity, denoted by "A," (2) the hours in which the unit operated at less than full capacity, due to mechanical outages, denoted by "B," and (3) the hours in which the unit operated at less than full capacity due to load-follow (economy) conditions, denoted by "C." So we have

$$\text{EA} = \frac{\text{A} + \text{B} + \text{C} - \text{B} + \text{EB}}{\text{PH}}, \text{ or}$$

(Equation C5)

$$\text{EA} = \frac{\text{A} + \text{EB} + \text{C}}{\text{PH}}.$$

Now consider capacity factor. Since generation by a unit is the sum of generation during hours of full operation and of partial operation, we have

(Equation C6)

$$\text{CF} = \frac{\text{A} + \text{EB} + \text{EC}}{\text{PH}},$$

A = hours of full capacity operation

EB = equivalized operation (in hours) during hours of partial operation due to mechanical outages

EC = equivalized operation (in hours) during hours of partial operation due to economy conditions.

Then, comparing Equations C 5 and C 6, we have

(Equation C7)

$$\text{EA} - \text{CF} = \frac{\text{C} - \text{EC}}{\text{PH}}$$

Equation C7 may be rewritten as

(Equation C8)

$$\text{C} - \text{EC} = (\text{EA} - \text{CF}) \times \text{PH}$$

Furthermore, EPEOH—the equivalized value (in hours) of partial economy outage hours—is equal to C − EC. That is, the equivalized value of partial economy outages equals the total number of hours in which the unit experienced partial economy outages, minus the equivalent operation during such hours. We therefore have

(Equation C9)

$$C - EC = EPEOH.$$

Comparing Equations C8 and C9 gives the relationship for equivalized partial economy outage hours,

$$EPEOH = (EA - CF) \times PH,$$

which we wrote as Equation C4.

The value of optimal capacity factor, which was shown to equal 1 minus EEOH/PH in Equation C2, is thus given by

$$OCF = 1 - \frac{FEOH + EPEOH}{PH},$$

from Equation C3, or, using Equation C4,

(Equation C10)

$$OCF = 1 - \frac{FEOH + (EA - CF) \times PH}{PH}$$

5. Estimating Optimal Capacity Factor in this Study

From Equations C1 and C10, a unit's capacity performance is defined as the ratio of capacity factor to optimal capacity factor, while the unit's optimal capacity factor is a function of four parameters: full economy outage hours, equivalent availability, capacity factor, and the number of hours in the period in question. Capacity performance can then be computed from the values of these four parameters.

The Edison Electric Institute's Equipment Availability data base compiles the values of these parameters, along with other extensive data reported by utilities to EEI, for most U.S. generating units on an annual basis. Full access to the EEI data base would thus enable one to compute the capacity performance of most U.S. generating units.

However, EEI does not permit users of the data base to obtain data on an individual unit basis.* Our access to the EEI data was therefore limited to obtaining data on an aggregate basis, so that we have obtained average values of the four parameters—FEOH, equivalent availability, capacity factor, and period hours—for each separate year of operation during the period 1968 through 1973, for each capacity size group 100-199 Mw, 200-299 Mw, up

*This policy was adopted to encourage full data reporting by utilities by ensuring confidentiality of information.

through 1000+ Mw.* (EEI considers 1968 to be the first year in which equivalent availability was accurately reported by utilities, and in fact the number of units for which this parameter was reported is several times greater starting in 1968 than in previous years.)

From the values of the four parameters in the EEI data base we have computed values of optimal capacity factor for each size group in each year as indicated in Table C1.

Because of the generally low variability from year to year in the average values of optimal capacity factor for each size group, it was decided to apply the 1968-1973 average value of optimal capacity factor from Table C1 in computing the values of capacity performance used in this study for all unit-years (*i.e.*, the period of operation 1961-1973)** The average values of capacity factor, optimal capacity factor, and capacity performance for each coal-fired size group are summarized in Table C2.

*This data was obtained from the EEI data base under a "Request for Special Analysis," and is not contained in the annual Report on Equipment Availability issued by EEI.

**Regression equations were developed relating the average annual optimal capacity factor to each year of operation, for each size group. The equations indicate slight decreases in optimal capacity factor over time for each size group under 500 Mw (*i.e.*, increases in load-following), and slight increases in optimal capacity factor over time for 500-599 Mw and 600+ Mw. It was decided to use the average value of optimal capacity factor rather than annual values from the regression equations, for simplicity and to apply a uniform measure of optimal capacity factor in computing capacity performance values for unit-years prior to 1968.

A regression equation was also obtained for the 1968-1973 average values of optimal capacity factor as a function of the size group. The equation is, Optimal Capacity Factor = 88.14 + 0.77 × B, where B = 0 for the 100-199 Mw group, 1 for the 200-299 Mw group, up through 5 for the 600+ Mw group. The equation indicates that average optimal capacity factor generally increases as a function of unit size, *i.e.*, larger units undergo less load-following. This result is in accord with perceived utility experience.

TABLE C1: OPTIMAL CAPACITY FACTOR FROM EEI DATA†

(OCF%/Number of Units)

Year of Operation	100-199 Mw	200-299 Mw	300-399 Mw	400-499 Mw	500-599 Mw	600+ Mw
1968	86.9/148	93.3/62	93.8/24	91.1/8	85.8/6	88.5/8
1969	89.4/148	94.1/59	93.8/24	88.7/8	92.3/16	93.5/8
1970	86.3/147	91.0/62	92.6/26	85.6/8	90.2/19	93.5/19
1971	83.2/159	86.9/61	89.0/28	86.8/10	90.5/24	94.5/19
1972	87.6/169	89.4/60	95.2/28	88.6/12	93.5/24	94.5/22
1973	86.0/144	88.4/43	89.2/24	84.2/12	88.6/15	90.3/23
1968-73	86.55/915	90.61/347	92.27/154	87.35/58	90.87/104	92.78/99

†The following procedures were followed in developing Table C1:

1. Data was drawn from units identified by EEI as "Coal-fired Only."
2. All data for units 600 Mw and larger was grouped together, since only 29 unit-years were reported above 700 Mw.
3. Data was disregarded for 800-899 Mw units in 1968 and 1969 (two units each year) and for 1000+ Mw units in 1971 and 1972 (one unit each year), since average capacity factor was reported to be greater than average equivalent availability for these cases, in apparent contradiction of the definitions of these parameters.
4. Average values of optimal capacity factor in the bottom line of Table C1 were computed by weighting the values of optimal capacity factor for each year by the total period hours for each year.
5. The average value of capacity factor used to compute optimal capacity factor for Table C1, using Equation C10, was taken from the EEI data rather than the CEP data base, in order to remain consistent with the use of EEI data in estimating the other parameters used to compute optimal capacity factor. The result is a conservative (low) estimate of capacity performance, since optimal capacity factor varies directly with capacity factor (see Equation C10), and EEI's definition of capacity factor results in higher average values of capacity factor than CEP's definition.
6. Full economy outage hours were computed as the difference between available hours and service hours.
7. Equivalized partial economy outage hours generally accounted for about 85% of total equivalent economy outage hours (vs. 15% for full economy outage hours), except for 600+ Mw units, for which equivalized partial economy outage hours accounted for only a little over half of all equivalent economy outage hours.

TABLE C2: AVERAGE CAPACITY PARAMETERS FOR COAL-FIRED UNITS

Size Group	Number of Unit Years	Average Capacity Factor	Average Optimal Capacity Factor	Average Capacity Performance
100-199 Mw	580	72.1%	86.55%	83.3%
200-299 Mw	420	69.0%	90.61%	76.2%
300-399 Mw	260	65.0%	92.27%	70.4%
400-499 Mw	76	57.1%	87.35%	65.4%
500-599 Mw	131	57.9%	90.87%	63.7%
600+ Mw	174	57.9%	92.78%	62.4%

Appendix D: Statistical Analysis of Nuclear Capacity Factors Through June 30, 1976

This Appendix presents the statistical analysis of nuclear plant capacity factors during the 6-month period from January 1, 1976, through June 30, 1976. The data base for this period includes the 38 nuclear units in the CEP data base which entered commercial service before 1975 and 10 new nuclear units which went commercial that year. Capacity factors are presented for the 6-month period and for the full commercial history of each reactor through mid-1976. In addition, statistical trend equations relating capacity factor to unit size and age from Chapter 3 are updated using seasonally adjusted 1976 data.

Following are the major results of this analysis:

1. The average 1976 6-month capacity factor of the 48 nuclear units was 52.2%. This is equivalent to 56.4%, seasonally correcting for refueling. The cumulative average capacity factor of the 48 units through mid-1976 was 58.1%, down from 59.3% at the end of 1975.

2. All six nuclear units with the highest lifetime capacity factors through mid-1976 had 6-month 1976 capacity factors equal to or greater than their lifetime average. Conversely, none of the seven reactors with the lowest lifetime capacity factors improved their capacity factors in the first half of 1976.

3. For nuclear plants of over 700 Mw capacity factor trend equations using data through mid-1976 indicate lower capacity factors than the trend equations through 1975.

1. Methodology

a. *1976 Capacity Factors:* Capacity factors for 1976 operation were computed by dividing total generation for the first six months of 1976 (as reported in the July, 1976 NRC Gray Book) by the product of the unit design electrical rating and the number of hours in the first six months of 1976.*

*Number of hours = 4367, which is 182 days × 24 hours per day minus 1 hour lost in transition to Daylight Savings Time in April.

TABLE D1: 1976 6-MONTH NUCLEAR CAPACITY FACTOR DATA

Sample	Number of Units	Average Capacity Factor	Standard Deviation
Units Commercial before 1975	38	55.6%	22.8%
Units Commercial in 1975	10	39.5%	25.8%
Units Commercial before 1976	48	52.2%	24.1%

b. *Lifetime Capacity Factors through June 30, 1976:* Lifetime reactor capacity factors were computed by averaging the capacity factors from each calendar year of commercial operation, including 1976, and weighting the 1976 capacity factor by 0.4985—the ratio of the hours in the first six months of 1976 to the number of hours in a normal year (8760). The lifetime capacity factor figures are thus actual capacity factors through June 30, 1976.

c. *Regression Analysis:* Generation data for the first six months of 1976 was extrapolated to all of 1976 in order to develop hypothetical whole-year data for the regression analyses. This was done by multiplying actual 6-month data by a factor of 2.158. This factor is the product of (i) the ratio of the hours in all of 1976 to the hours in the first six months of 1976 (2.0114); and (ii) the anticipated ratio of the average capacity factor for all of 1976 to the average capacity factor for the first six months of 1976 (1.073).

The latter figure is the computed ratio of the average 1975 capacity factor to the average in the first six months of 1975 for the 38 nuclear units in the CEP data base. It was applied to the 6-month 1976 data to reflect the anticipated higher average capacity factors in the last six months of 1976 as compared to the first six months, due to (i) lower frequency of scheduled refueling outages in the second half of the year and (ii) general increase in nuclear capacity factors with increasing unit age, attributable to the statistically significant maturation observed in PWR's.

The hypothetical 1976 generation figures computed by this procedure were then divided by the product of unit capacity and 8784 (total hours in 1976) to obtain hypothetical 1976 capacity factors from 6-month generation data of that year.

The regression equations thereby give double weight to the 1976 6-month data, and in the process adjust the data in anticipation of higher capacity factors during the second half of 1976. This procedure thus exaggerates the effect of the presence or absence of refueling outages in the first half of the year. However, it is probably a safe assumption that the impact of unit size and unit age—the independent variables tested in the regressions—is not distorted by this procedure in that there is probably little or no correlation between the scheduling of refueling outages in the calendar year and either unit size or age.

2. 1976 Capacity Factor Statistics

The 1976 6-month capacity factor data is summarized in Table D1. The average 52.2% capacity factor for the 48 commercial U.S. nuclear plants in the first half of 1976 is consistent with an entire-year capacity factor of 56.4% (applying the ratio of second-half to first-half capacity factors from 1975). This is less than the 60.4% actual average 1975 capacity factor for the 38 units in the CEP data base. While the average capacity of the 48 units (774 Mw) is greater than that of the 38 units (744 Mw), only 10 of the 48 units operating in 1976 are first-year units, compared to 12 new units of 38 in 1975. Thus, not only have 1976 nuclear plant capacity factors to date been less than capacity factors in 1975 and in most previous years, but it also appears that, holding constant the key variables of unit size and age, the 1976 performance has thus far been poorer than that of 1975.*

3. Lifetime Capacity Factor Statistics, Including 1976 Data

a. *Capacity Factor by Unit:* The lifetime capacity factors of the 38 nuclear units in the CEP data base, through June 30, 1976, are listed in Table D2 in decreasing capacity factor order and compared to lifetime capacity factors through the end of 1975.

Significantly, the six reactors with the highest lifetime capacity factors as of mid-1976 are all "mature" units (with over three years of commercial service), all of which operated in the first half of 1976 at capacity factors equal to or higher than their lifetime average. This indicates that nuclear units with a history of high capacity factors through 1975 continued their good performance in the first six months of 1976. Conversely, the seven reactors with the lowest mid-1976 lifetime capacity factors all had capacity factors in 1976 which were lower than their previous lifetime average, indicating that reactors with low lifetime capacity factors did not improve upon their performance in the first half of 1976 (five of the seven units have not yet operated for three full calendar years). Of the other 25 reactors in the data base, 12 had capacity factor increases, 12 had decreases, and one stayed virtually constant. The average lifetime capacity factor of the 38 units declined slightly, from 59.3% (110 unit-years) to 58.7% (129 unit-years).

The 1976 half-year capacity factors (which are also the lifetime capacity factors) of the 10 nuclear units which went commercial in 1975 range from 85.8% (Calvert Cliffs 1) to zero (Browns Ferry 2), and average 39.5%. They are listed by unit in the nuclear data base listing in Appendix A. Including these 10 new units, the average lifetime capacity factor of commercial U.S. nuclear units through June 30, 1976 is 58.1%.

b. *Capacity Factor by Reactor Vendor and Type:* Average capacity factors through mid-1976 for nuclear plants manufactured by the four reactor vendors,

*Of course, it is possible that more of the year's refueling outages were scheduled for the first half of the year in 1976 than in 1975, which could cause 1976 6-month capacity factors to be a greater underestimate of 1976 full-year data than was the case for 1975 data. Moreover, load-following by nuclear units may be increasing, which could cause 1976 capacity factor data to understate capacity performance relative to prior years. In particular, the low (10.4%) capacity factor of Portland General Electric's 1130-Mw Trojan nuclear plant in the first six months of 1976 may have resulted in part from load-following permitted by full reservoir capacity in the hydro-rich Pacific Northwest in 1976. However, it is unlikely that load-following has accounted for more than 1-2 percentage points of reduced average nuclear capacity factors in 1976.

TABLE D2: LIFETIME NUCLEAR CAPACITY FACTORS, THROUGH 6/30/76

Unit	Capacity	Capacity Factor through 12/31/75	Capacity Factor through 6/30/76	Number of Full Calendar Years
1. Point Beach 2	497 Mw	75.9%	75.9%	3.5
2. Connecticut Yankee	575 Mw	73.3%	73.3%	8.5
3. Robinson 2	707 Mw	71.0%	73.1%	4.5
4. Point Beach 1	497 Mw	69.0%	70.9%	5.5
5. San Onofre	450 Mw	68.8%	70.0%	8.5
6. Monticello†	545 Mw	66.2%	68.1%	4.5
7. Oconee 3	886 Mw	64.9%	66.7%	1.5
8. Prairie Island 2	530 Mw	68.4%	65.7%	1.5
9. Three Mile Island	819 Mw	77.2%	65.7%	1.5
10. Oyster Creek†	650 Mw	64.5%	63.2%	6.5
11. Kewaunee	560 Mw	68.1%	61.7%	1.5
12. Turkey Point 4	745 Mw	63.5%	61.4%	2.5
13. Ginna	490 Mw	63.4%	61.0%	5.5
14. Vermont Yankee†	514 Mw	58.2%	60.5%	3.5
15. Turkey Point 3	745 Mw	57.7%	60.1%	3.5
16. Maine Yankee	790 Mw	55.0%	59.8%	3.5
17. Quad Cities 2†	809 Mw	57.2%	59.7%	3.5
18. Cooper†	778 Mw	56.5%	57.4%	1.5
19. Nine Mile Point 1†	610 Mw	55.2%	56.5%	6.5
20. Oconee 2	886 Mw	64.0%	56.4%	1.5
21. Fort Calhoun	457 Mw	52.0%	56.3%	1.5
22. Quad Cities 1†	809 Mw	59.9%	56.1%	3.5
23. Millstone 1†	690 Mw	53.4%	55.3%	5.5
24. Peach Bottom 3†	1065 Mw	54.5%	55.3%	1.5
25. Oconee 1	886 Mw	59.8%	55.2%	2.5
26. Prairie Island 1	530 Mw	55.2%	54.5%	2.5
27. Arkansas 1	850 Mw	65.5%	54.3%	1.5
28. Surry 1	823 Mw	49.5%	53.9%	3.5
29. Surry 2	823 Mw	53.3%	53.4%	2.5
30. Dresden 2†	809 Mw	53.4%	52.0%	3.5
31. Dresden 3†	809 Mw	49.4%	51.1%	4.5
32. Indian Point 2	873 Mw	53.7%	50.9%	2.5
33. Peach Bottom 2†	1065 Mw	54.5%	50.4%	1.5
34. Zion 2	1050 Mw	52.5%	48.3%	1.5
35. Pilgrim 1†	670 Mw	49.0%	43.7%	3.5
36. Zion 1	1050 Mw	45.6%	42.2%	2.5
37. Palisades	821 Mw	23.2%	22.5%	4.5
38. Browns Ferry 1†	1098 Mw	14.0%	9.4%	1.5
38 Units		59.3%	58.7%	127

†Indicates BWR, all others are PWR.

and divided by reactor type, are in Table D3.

The average capacity factors of Westinghouse and GE reactors declined slightly in the first half of 1976. The six Babcock & Wilcox reactors declined in average capacity factor from 65.2% through the end of 1975 to 56.8% through mid-1976, due largely to refueling at three units and the poor 1976 capacity factor of the new Rancho Seco unit (18.0%). In contrast, the average cumulative capacity factor of Combustion Engineering reactors increased from 38.8% to 44.2%, largely from very high capacity factors during the first half of 1976 at the new Calvert Cliffs unit (85.8%) and Maine Yankee (88.1%). Absent Palisades, Combustion Engineering's first reactor, with a cumulative 22.5% capacity factor, Combustion Engineering reactors have an average capacity factor through mid-1976 of 60.4%, comparable to the general level of Westinghouse and B&W reactors.

4. Regression Equations Through Mid-1976

In Section 1c of this Appendix we described how we extrapolated 1976 half-year generation to obtain hypothetical full-year 1976 capacity factors to use with pre-1976 data in statistical regression analysis. The key equations resulting from this analysis are set forth below and are compared to equations developed in Chapter 3 based on the CEP nuclear data base through 1975.

a. *PWR Regression Equations*

Through mid-1976:

(Equation D1)*

$$\text{Capacity Factor} = 97.79 - \underset{(1.018)}{4.013} \times (\text{Mw}/100) - \underset{(2.32)}{4.18} \times \text{AGED}$$

$$N = 95, \text{ Adjusted } R^2 = 0.22$$

Through 1975:

(Equation 3.1)

$$\text{Capacity Factor} = 93.48 - \underset{(1.140)}{3.368} \times (\text{Mw}/100) - \underset{(2.36)}{4.22} \times \text{AGED}$$

$$N = 66, \text{ Adjusted } R^2 = 0.21$$

The ten-year levelized average versions of the two equations are:

(Equation D2)

Through mid-1976: $\text{Capacity Factor} = 90.74 - 4.013 \times (\text{Mw}/100)$

(Equation 3.2)

Through 1975: $\text{Capacity Factor} = 86.38 - 3.368 \times (\text{Mw}/100)$

*The error coefficients for the new equations are given in parentheses below the regression coefficients. The "AGED" term in the equations is a function of the age of the unit, and is defined: AGED = 1/Log (1 + AGE).

Equation D2 reflects half-year 1976 PWR experience along with experience through 1975. It has a higher intercept but a steeper slope (coefficient of Mw/100) than Equation 3.2, indicating a greater inverse relationship between unit size and PWR capacity factor than is in evidence through the end of 1975. The cross-over point for the two equations is 676 Mw. Below that unit size PWR performance would be expected to improve on the basis of data from the first half of 1976. Above 676 Mw PWR capacity factors would be expected to decline. This reflects the generally low half-year 1976 capacity factors of larger PWR's, especially 1050-Mw Zion 1 and Zion 2, the new 1130-Mw Trojan reactor, and most of the Babcock & Wilcox units.

The similarity in the AGED coefficients in Equations D1 and 3.1 indicates that 1976 data has not affected the extent of the general maturation tendency observed for PWR's.*

b. *BWR Regression Equations*

(Equation D3)

Through mid-1976: $$\text{Capacity Factor} = 93.64 - \underset{(1.350)}{5.343} \times (\text{Mw}/100)$$

$$N = 63, \text{ Adjusted } R^2 = 0.20$$

(Equation 3.3)

Through 1975: $$\text{Capacity Factor} = 79.30 - \underset{(1.461)}{3.287} \times (\text{Mw}/100)$$

$$N = 44, \text{ Adjusted } R^2 = 0.11$$

Equation D3 includes BWR operation through mid-1976. It has a higher intercept but far more negative slope than Equation 3.3, indicating a much greater inverse relationship between capacity factor and unit size for BWR's than is reflected for experience through the end of 1975 in Equation 3.3. The increase in the Adjusted R^2 value from 0.11 to 0.20 denotes an increase in the extent to which variation in BWR capacity factors is correlated with variations in the size (capacity) of the units.

The cross-over value for Equations D3 and 3.3 is 697 Mw. Below that unit size higher capacity factors are indicated by Equation D3 than by Equation 3.3. Above 697 Mw Equation D3 yields significantly lower capacity factors, reflecting the zero 1976 capacity factors to date of Browns Ferry 1 and 2 (1098 Mw each) which were still being repaired for damage suffered in the March

*In testing for the effect of unit duplication on PWR capacity factors through mid-1976 we obtained the relationship, Capacity Factor = 98.33 − 4.132 × (Mw/100) − 4.645 × AGED + 6.86 × DUPE, where DUPE= 1 for a duplicate unit and 0 otherwise (see Section 5.5). However, the error coefficient of the DUPE term is 4.62, so that the coefficient is statistically significant at only the 93% confidence level, or less than the 95% level used as a cut-off point in this study. In contrast, the DUPE term in Equation 5.1 (which reflects PWR experience through the end of 1975) is significant at the 97% level. While the lesser statistical significance of DUPE in the present equation may result in part from more frequent refueling by duplicate units in early 1976, it does indicate the tentativeness of our finding that duplicate PWR's have higher capacity factors than other PWR's. Moreover, the size of the DUPE term itself is reduced from 10 percentage points in Equation 5.1 to under 7% here.

TABLE D3: NUCLEAR CAPACITY FACTORS BY REACTOR VENDOR AND TYPE

Reactor Manufacturer	Through 12/31/75 Number of Units	Through 12/31/75 Number of Unit Years	Through 12/31/75 Average Capacity Factor	Through 6/30/76 Number of Units	Through 6/30/76 Number of Unit Years	Through 6/30/76 Average Capacity Factor
Babcock & Wilcox	5	6	65.2%	6	9	56.8%
Combustion Engineering	3	8	38.8%	5	10.5	44.2%
Westinghouse	16	52	64.7%	18	61	63.8%
All PWR's	24	66	61.6%	29	80.5	60.5%
General Electric (BWR's)	14	44	55.9%	19	53.5	54.4%
All Nuclear Units	38	110	59.3%	48	134	58.1%

TABLE D4: LEVELIZED 10-YEAR NUCLEAR CAPACITY FACTORS FROM REGRESSION EQUATIONS

	PWR's		BWR's	
Unit Size	Based on Operation Through 12/31/75	Based on Operation Through 6/30/76	Based on Operation Through 12/31/75	Based on Operation Through 6/30/76
600 Mw	66.2%	66.7%	59.6%	61.6%
800 Mw	59.4%	58.6%	53.0%	50.9%
1000 Mw	52.7%	50.6%	46.4%	40.2%
1150 Mw	47.6%	44.6%	41.5%	32.2%

1975 cable tray fire.*

c. Comparison of Capacity Factor Values from the Regression Equations

Table D4 gives 10-year levelized average capacity factors for BWR's and PWR's using Equations D2 and D3 which reflect mid-1976 operation. These values are compared to those obtained using Equations 3.2 and 3.3 based on operation through 1975.

The capacity factor figures in Table D4 should be accepted with caution, particularly values for large BWR's, since they are derived from equations which exaggerate effect of operation during the first half of 1976 (Section 1)—especially outages at Browns Ferry 1 and 2. However, the figures, and new regression equations overall, indicate that operating data at nuclear units in the first half of 1976 reinforce the conclusion that nuclear capacity factors are inversely related to unit size.

*As was the case for data through the end of 1975, unit age is not correlated with capacity factor for BWR operation through mid-1976. In the relationship, Capacity Factor = 94.56 − 5.192 × (Mw/100) − 0.94 × AGED for BWR's, which includes 1976 data, the error coefficient of AGED is 2.73. The ratio of coefficient to error coefficient is so small that the AGED coefficient is not even significant at the 65% confidence level.

Appendix E: Nuclear Capacity Factor Regression Equations

The statistical regression equations developed in this study for nuclear capacity factors are listed below. Equations appearing in the text are identified. Standard error coefficients are given in parentheses below the coefficients. Equations whose coefficients are not significant at the 95% confidence level are noted with an asterisk. Correlation coefficient matrices are given at the end of this Appendix.

The following abbreviations are used:

CF denotes Capacity Factor

AGED is a function of unit age, equal to 1/Log (1 + AGE)

Mw denotes megawatt capacity

CODA denotes initial date of commercial operation (last two digits of year)

DUPE equals one for duplicate units, zero otherwise

PROT equals one for the prototype unit of a duplicate pair, zero otherwise

CHIC equals one for Commonwealth Edison units, zero otherwise

PWR Operation Through 1975

(Equation 5.1)

$$CF = 95.45 - \underset{(1.117)}{3.502} \times (Mw/100) - \underset{(2.39)}{5.43} \times AGED$$

$$+ \underset{(5.10)}{10.00} \times DUPE$$

N = 66

Adjusted $R^2 = 0.25$

(Equation 3.1)

$$CF = 93.48 - 3.368 \times (Mw/100) - 4.22 \times AGED$$
$$\qquad\qquad (1.140) \qquad\qquad\qquad (2.36)$$

$$N = 66$$
$$\text{Adjusted } R^2 = 0.21$$

$$CF = 88.81 - 4.116 \times (Mw/100)$$
$$\qquad (1.079)$$

$$N = 66$$
$$\text{Adjusted } R^2 = 0.19$$

$$*CF = 95.57 - 3.545 \times (Mw/100) - 5.45 \times AGED$$
$$\qquad\qquad (1.147) \qquad\qquad\qquad (2.42)$$

$$+ 10.25 \times DUPE + 0.73 \times PROT$$
$$(5.41) \qquad\qquad (4.46)$$

$$N = 66$$
$$\text{Adjusted } R^2 = 0.23$$

$$*CF = -14.24 - 4.282 \times (Mw/100) - 6.01 \times AGED$$
$$\qquad\qquad (1.336) \qquad\qquad\qquad (2.73)$$

$$+ 1.66 \times CODA$$
$$(1.28)$$

$$N = 66$$
$$\text{Adjusted } R^2 = 0.22$$

PWR's through 1975, excluding Connecticut Yankee

$$*CF = 89.38 - 3.311 \times (Mw/100) - 3.01 \times AGED$$
$$\qquad\qquad (1.165) \qquad\qquad\qquad (2.60)$$

$$N = 58$$
$$\text{Adjusted } R^2 = 0.18$$

$$CF = 85.51 - \underset{(1.094)}{3.791} \times (Mw/100)$$

$N = 58$
Adjusted $R^2 = 0.18$.

PWR's through 1975, excluding Palisades

$$CF = 90.00 - \underset{(0.933)}{1.769} \times (Mw/100) - \underset{(1.93)}{6.31} \times AGED$$

$N = 62$
Adjusted $R^2 = 0.27$

PWR's through 1975, Age = 1

$$*CF = 65.61 - \underset{(1.723)}{1.432} \times (Mw/100)$$

$N = 24$
Adjusted $R^2 = 0.03$

PWR's through 1975, Age = 2

$$CF = 82.19 - \underset{(1.544)}{2.989} \times (Mw/100)$$

$N = 16$
Adjusted $R^2 = 0.21$

PWR's through 1975, Age = 3 or greater

$$CF = 113.47 - \underset{(2.467)}{7.920} \times (Mw/100)$$

$N = 26$
Adjusted $R^2 = 0.30$

PWR's in 1973

$$*CF = 95.16 - \underset{(2.315)}{8.170} \times (Mw/100) + \underset{(3.84)}{5.58} \times AGED$$

N = 10
Adjusted $R^2 = 0.61$

PWR's in 1974

$$*CF = 98.70 - \underset{(3.653)}{4.891} \times (Mw/100) - \underset{(7.41)}{4.58} \times AGED$$

N = 16
Adjusted $R^2 = 0.23$

PWR's in 1975

$$*CF = 88.66 - \underset{(1.219)}{2.989} \times (Mw/100) - \underset{(2.66)}{0.47} \times AGED$$

N = 24
Adjusted $R^2 = 0.21$

PWR's through mid-1976, using half-year seasonally adjusted 1976 data to represent all of 1976

(Equation D.1)

$$CF = 97.79 - \underset{(1.018)}{4.013} \times (Mw/100) - \underset{(2.32)}{4.18} \times AGED$$

N = 95
Adjusted $R^2 = 0.22$

$$*CF = 98.33 - \underset{(1.015)}{4.132} \times (Mw/100) - \underset{(2.33)}{4.65} \times AGED + \underset{(4.62)}{6.86} \times DUPE$$

N = 95
Adjusted $R^2 = 0.23$

BWR Operation Through 1975

(Equation 3.3)

$$CF = 79.30 - 3.287 \times (Mw/100)$$
$$(1.461)$$

$$N = 44$$
$$\text{Adjusted } R^2 = 0.11$$

$$*CF = 83.33 - 3.562 \times (Mw/100) - 0.82 \times AGE$$
$$(1.558) \qquad (1.50)$$

$$N = 44$$
$$\text{Adjusted } R^2 = 0.09$$

$$*CF = 85.46 - 3.218 \times (Mw/100) - 0.09 \times CODA$$
$$(1.898) \qquad (1.63)$$

$$N = 44$$
$$\text{Adjusted } R^2 = 0.08$$

$$*CF = 80.07 - 4.536 \times (Mw/100) + 3.12 \times AGED + 4.04 \times CHIC$$
$$(1.777) \qquad (2.74) \qquad (5.06)$$

$$N = 44$$
$$\text{Adjusted } R^2 = 0.10$$

BWR's through 1975, AGE = 1

$$*CF = 74.58 - 2.234 \times (Mw/100)$$
$$(2.624)$$

$$N = 14$$
$$\text{Adjusted } R^2 = 0.06$$

BWR's through 1975, AGE = 2

$$*CF = 71.32 - \underset{(3.053)}{2.433} \times (Mw/100)$$

N = 10
Adjusted R^2 = 0.07

BWR's through 1975, AGE = 3 or greater

$$CF = 120.38 - \underset{(2.530)}{9.614} \times (Mw/100)$$

N = 20
Adjusted R^2 = 0.45

BWR's in 1973

$$*CF = 29.85 + \underset{(4.642)}{2.986} \times (Mw/100) + \underset{(6.14)}{3.88} \times AGED$$

N = 10
Adjusted R^2 = 0.04

BWR's in 1974

$$*CF = 87.57 - \underset{(2.868)}{1.269} \times (Mw/100) - \underset{(9.19)}{13.81} \times AGED$$

N = 10
Adjusted R^2 = 0.23

BWR's in 1975

$$*CF = 96.66 - \underset{(3.249)}{8.461} \times (Mw/100) + \underset{(6.94)}{9.98} \times AGED$$

N = 14
Adjusted R^2 = 0.37

BWR's through mid-1976, using half-year seasonally adjusted 1976 data to represent all of 1976

(Equation D3)

$$CF = 93.64 - 5.343 \times (Mw/100)$$
$$(1.350)$$

$$N = 63$$
$$\text{Adjusted } R^2 = 0.20$$

$$^*CF = 94.56 - 5.192 \times (Mw/100) - 0.94 \times AGED$$
$$(1.428) \qquad\qquad (2.73)$$

$$N = 63$$
$$\text{Adjusted } R^2 = 0.19$$

Correlation Coefficient Matrices

PWR – 66 Unit Years

	Mw	CODA	AGE
Mw	—	0.61	−0.42
CODA	0.61	—	−0.66
AGE	−0.42	−0.66	—

BWR – 44 Unit Years

	Mw	CODA	AGE
Mw	—	0.63	−0.32
CODA	0.63	—	−0.51
AGE	−0.32	−0.51	—

Appendix F: Statistical Concepts and Terms

1. Terms in Regression Analysis

Regression analysis is a statistical technique for estimating the extent to which changes in the value of a dependent variable are related to changes in one or more independent variables. The dependent variable of concern in this study is capacity factor (capacity performance, for coal-fired units). Independent variables tested for their statistical significance to capacity factor or capacity performance are unit age, size, and date of installation for nuclear plants; and age, size, period of installation, coal sulfur content, coal ash content, and boiler criticality for coal plants.

The statistical significance of a regression equation may be indicated by (1) the t-ratios of the correlation coefficients and (2) the multiple correlation coefficients (adjusted R-squared value).

(1) *T-ratios:* The t-ratio is the ratio of a correlation coefficient to its error coefficient. For example, in Equation 2.1 (see Chapter 2 or Appendix E), the correlation coefficient for PWR unit size is 3.368, while the error coefficient, given directly below the correlation coefficient, is 1.140. Their ratio, equal to 2.95, is called the t-ratio. The lower the t-ratio, the greater is the probability that the actual relationship between dependent and independent variable (PWR capacity factor and unit size) deviates from the relationship indicated by the coefficient (3.368). Therefore, the t-ratio of a correlation coefficient determines the statistical confidence which may be placed in that coefficient.

Confidence values for different values of t-ratios may be determined by consulting a table of "Student's t-test" in any statistics textbook. A 95% confidence limit has been used in this study (equal to 90% for a two-tailed distribution) as the threshold of statistical significance. Regression equations with one or more correlation coefficients not significant to the 95% level have not been accepted here as statistically significant.

By consulting a Student's t-test table, the t-ratio of 2.95 for PWR unit size in Equation 2.1 is seen to be significant at the 99.5% level. The coefficient for

AGED in the same equation, with a t-ratio of 1.79, is significant only to the 96% level which, nevertheless is greater than the 95% threshold. Where different functional forms of variables have been tested in regression equations in this study, we have used the form with the greatest significance, as indicated by t-ratios.

(2) *Adjusted R-squared:* The adjusted R^2 value of a regression equation, or multiple correlation coefficient, gives the correlation between the actual values of a dependent variable (capacity factor) and the values of that variable obtained from the regression equation. The adjusted R^2 thus measures the goodness of fit of the regression equation to the data from which the equation has been drawn. The adjusted R^2 value may range from 0 to 1, with higher values indicating a good fit of equation to data.

Adjusted R^2 values for equations presented in this study range from 0.03 to 0.41. For the key equations used to estimate capacity factor or capacity performance as a function of unit size, the adjusted R^2 ranges from 0.11 to 0.41. These values are less than those usually found in statistical research. However, the low values in this study are attributable in part to the use of a calendar-year measuring framework, which magnifies the effect of capacity factor variations at individual units. This is especially true for nuclear plants, which frequently refuel twice in one year and not at all in the next, thereby causing significant year-to-year capacity factor variation which is inherently erratic and not explainable by variations in unit age, size, or date of installation.

2. Levelized Average Capacity Factors

A levelized average is a technique for accounting for the time value of money in averaging a series of costs which vary over time. It weights more heavily dollars received or spent in early years, since the present worth of those dollars is greater than for dollars received or spent further in the future. The rate at which the value of dollars depreciates from year to year is measured by a discount rate measuring the cost of money. This is usually taken to be the interest rate on borrowed capital. Cost levelizing is standard utility practice in projecting power plant generating costs.

Capacity factors represent kilowatt-hour generation, and are not costs *per se*. However, they affect generating costs, since plant per-kwh capital costs are inversely proportional to kwh's generated. In effect, capacity factors in the initial years of plant operation are more significant than later capacity factors, since they determine initial per-kwh costs. Utilities which anticipate changes in capacity factor over time are therefore increasingly computing projected average lifetime capacity factors for new plants on a levelized basis (for example, see LILCO reference in the Bibliography, Part 5).

This study projects 10-year rather than 30-year capacity factors as projected by many utilities, because the capacity factor data base extends to only about ten years of operation (eight for commercial nuclear units, 13 for coal). A discount rate of 10% is used, since this is currently the average cost of money for electric utilities. At this discount rate the present worth of generation produced in a given year is 1.1 times as great (10% greater) as generation in the following year. If the projected capacity factor in each year is denoted by CFi, where i is the year (so that CF_1 is the first-year capacity factor, CF_2 the second-year capacity factor, etc.), then the levelized average 10-year capacity factor is given by Equation F1.

(Equation F1)

$$\text{Levelized Capacity Factor} = \frac{CF_1 + CF_2/1.1 + CF_3/1.1^2 + \ldots + CF_{10}/1.1^9}{1 + 1/1.1 + 1/1.1^2 + \ldots + 1/1.1^9}$$

Since all capacity factor trend equations in this study include age as a significant variable, except the equation for BWR's, Equation F1 has been used to convert age-dependent capacity factors to 10-year levelized averages. For convenience in computation, we list below the 10-year levelized value of the age functions used in this study:

a. AGED: 1.688 (AGED = 1/Log (1 + AGE))
b. AGE: 4.725
c. $c_1XA + c_2\,AGE$: $c_1/6.76 + c_2 \times 4.725$
   (XA = 1 for AGE = 1, and 0 otherwise)

Levelized averages are less than conventional arithmetic averages for trend equations where capacity factor increases with increasing age (PWR's, supercritical coal, large coal) and are greater where capacity factor decreases with increasing age (subcritical coal under 600 Mw capacity). However, the difference is not great. For example, the projected levelized 10-year average PWR capacity factor from Equation 2.1 is only 0.73 percentage points less than the arithmetic average computed from the same equation.

Bibliography

Power Plant Capacity Factor Statistics

Comey, David D. "Nuclear Power Plant Reliability: The 1973-74 Record." *Not Man Apart*, (mid-April 1975). Originally published by Business and Professional People for the Public Interest (February 1975).

__________. "Will Idle Capacity Kill Nuclear Power." *Bulletin of The Atomic Scientists* (November 1974). Originally published by Business and Professional People for the Public Interest (September 1974).

Commonwealth Edison Co. Staff. "Appraisal of Nuclear Plant Reliability." *Power Engineering* (May 1975).

Edison Electric Institute, "Report on Equipment Availability for the Ten-Year Period, 1965-1974." EEI Publication No. 75-50 (December 1975). Published annually.

Kohn, Harold W. "Reactor Performance Evaluation." *Power Engineering* (December 1975).

Lee, Byron. "Economics of Availability." Commonwealth Edison Co. (October 1974).

Rossin, A. David. "On the Credibility of Arguments Concerning Nuclear Power Plant Reliability." Commonwealth Edison Co. (late 1974).

S.M. Stoller Corporation. "Survey of Operating Statistics of Light Water Reactors in the United States" (April 11, 1975).

__________. "Performance Characteristics of Nuclear Power Plants in the United States" (March 6, 1976).

U.S. Atomic Energy Commission. "Evaluation of Nuclear Power Plant Availability." Report #OOE-ES-001 (January 1974).

__________. "Nuclear Power Plant Availability and Capacity Statistics for 1973." Report #OOE-OS-002 (May 1974).

U.S. Nuclear Regulatory Commission. "Operating Units Status Report." (monthly, beginning July 1974). Also known as "The Gray Books."

Power Plant Reliability Overview

Cole, G.F. "Replication: One Way to Cost and Schedule Savings." *Power Engineering* (July 1976), pp. 44-49.

Mechanics Research,Inc. "Study on the Reliability of Large Coal-Fired and Nuclear Power Plants." (December 15, 1975).

Nuclear Services Corporation. "Productivity Improvement Action Plan." (November 28, 1975).

Trident Associates. "Power Plant Productivity—An Assessment of Outages at Two Coal-Fired and Two Nuclear Plants." (November 17, 1975).

U.S. Federal Energy Administration. "A Report on Improving the Productivity of Electric Power Plants." (March 1975).

U.S. Federal Power Commission. "Electrical Generating Plant Availability." Bureau of Power (May 1975).

Nuclear Plant Reliability

Billingham, R. "Fifteen Years in Nuclear Power." *Peaceful Uses of Atomic Energy* (Proceedings of the 1974 IAEA Conference, 1975).

Bridenbaugh, D.G., R.L. Turner and G.B. Lloyd. "Maintenance and In-Service Inspection Experience at Large Nuclear Power Plants." *Peaceful Uses of Atomic Energy, op. cit.*
Kerr, W., Chairman, Advisory Committee on Reactor Safeguards, Nuclear Regulatory Commission. Unpublished letter to William A. Anders, Chairman, NRC (February 8, 1975).
Kniazewycz, B.G. and W.C. McArthur. "Is Your In-Plant Exposure as Low as Practicable?" *Power Engineering* (August 1975).
Lapides, M.E., and E.L. Zebroski. "Use of Nuclear Plant Operating Experience to Guide Productivity Improvement Programs." Electric Power Research Institute, EPRI SR-26-R (November 1975).
Lapides, M.E., E.L. Zebroski and L. Johnson. "Developing Methods for Establishing Improved Plant Reliability." Paper presented at American Power Conference, April 20-22, 1976.
Minnick, Lawrence E. "Operating Experience with Light Water Reactors in the USA." Electric Power Research Institute (April 1975).
Muntzing, L. Manning. "Improving the Reliability of Nuclear Power Plants—A Point of View." Director of Regulation, Atomic Energy Commission. Unpublished memorandum (December 2, 1974).
Ottoson, H.L. and W.R. Gould. "Operating Experience with San Onofre Nuclear Generating Station." *Peaceful Uses of Atomic Energy, op. cit.*
U.S. Nuclear Regulatory Commission. "Occupational Radiation Exposure at Light Water Cooled Power Reactors, 1969-1974." Report #NUREG-75/032 (June 1975).
Verna, Bernard J. Bimonthly Column in *Nuclear News* (beginning in May 1975).

Fossil Plant Reliability

Graham, John. "Combustion Optimization." *Electrical World* (June 15, 1976).
Gray, R.J. and G.F. Moore. "Burning the Sub-Bituminous Coals of Montana and Wyoming in Large Utility Boilers." Paper presented at the Winter Annual Meeting of The Fuel Division of The American Society of Mechanical Engineers, November 17-22, 1974.
Komanoff, C., H. Miller and S. Noyes. *The Price of Power*. New York: Council on Economic Priorities, 1972.
Power Engineering Staff. "Supercritical Boilers: Confusion Abounds and the Trend is Down." *Power Engineering* (April 1971).
Thomas, E.F., J.G. Holmes, Jr. and C.C. Schonhoff. "Operating Experience with Large Supercritical Units." *Combustion* (April 1971).
Tillinghast, John and John E. Dolan. "AEP Succeeds With Large Units." *Electrical World* (August 1, 1976).

Power Generation Economics

Bupp, Irvin C., *et. al*. "Trends in Light Water Reactor Capital Costs in the United States: Causes and Consequences." Center for Policy Alternatives, M.I.T., CPA 74-8, December 18, 1974; abridged in *Technology Review* (February 1975).
Cagnetta, John. Testimony before the Connecticut Public Utilities Control Authority in Hearings on the Economics of Nuclear Power, Northeast Utilities (January 1976).
Comtois, Wilfred H. "Escalation, Interest During Construction and Power Plant Schedules." Prepared for Westinghouse Power Systems Marketing (September 1975).

Dubin-Mindell-Bloome Associates. "A Study of Electrical Energy Usage in the Public Service Electric & Gas Company Service Territory." (June 30, 1976).
Ebasco Services, Inc. "Fossil and Nuclear 1000 Mw Central Station Power Plants Investment Estimates." (September 1975).
Edison Electric Institute. "Nuclear Fuels Supply." (March 1976).
Harding, Jim. "Nuclear Fuel Burnup and Fuel Cycle Costs." California Energy Resources Conservation & Development Commission (June 1976).
Investor Responsibility Research Center. "The Nuclear Power Alternative." Special Report 1975-A (January 1975).
Lanoue, Ron. "Nuclear Plants: The More They Build, The More You Pay." Center for Study of Responsive Law (1976).
Lee, T.H. "Power Generation Economics." General Electric (January 22, 1976).
Arthur D. Little, Inc. "Report to Long Island Lighting Company on Base-Load Alternatives for the 1981-1984 Period." (June 1974).
__________. "Economic Comparison of Base-Load Generation Alternatives for New England Electric." (March 1975).
Long Island Lighting Company. "Supplement 3, Revised." Submitted in Case 80003, LILCO Application before the New York State Board on Electric Generation Siting and the Environment to construct two 1150-Mw nuclear units at Jamesport, Long Island (September 1974).
Massachusetts Energy Policy Office. "The Economics of Nuclear Power: A New England Perspective." (December 1975).
Reichle, Leonard F.C. "The Economics of Nuclear Power." Ebasco Services, Inc. (August 27, 1975).
Schultheis, Walter T. Testimony before the Connecticut Public Utilities Control Authority, in Hearings on the Economics of Nuclear Power, Northeast Utilities (January 1976).
U.S. Atomic Energy Commission. "Power Plant Capital Costs: Current Trends and Sensitivity to Economic Parameters." WASH-1345 (October 1974).
U.S. Energy Research and Development Administration. "CONCEPT— A Computer Code for Conceptual Cost Estimates of Steam-Electric Power Plants—Phase IV, User's Manual." ERDA-108 (June 1975).
U.S. Federal Power Commission. "Steam-Electric Plant Construction Cost and Annual Production Expenses." (1973 and previous annual issues).

CEP Reports on Nuclear and Coal Economics

Testimony for the New Mexico Attorney General, before the New Mexico Public Service Commission, on the proposed Palo Verde Nuclear Generating Station, August 4, 1976. Focuses on capacity factors and power costs for a Western U.S. site.

Economic Analysis of the Nuclear Expansion Program of Public Service Electric & Gas Company, before the New Jersey Public Utilities Commission, prepared for the New Jersey Department of Public Advocate, June 7, 1976. Most detailed CEP analysis of nuclear and coal costs, early installation of nuclear plants, and availability of Eastern low-sulfur coal.

Testimony for the Connecticut Public Utilities Control Authority, in Generic Hearings on the Economics of Nuclear Power, Feburary 27, 1976.

Testimony for the Riverhead First Committee, in New York Public Service Commission Case 80003, LILCO Application to Construct the Jamesport Nuclear Power Station, November 10, 1975.

"Responding to Con Edison: An Analysis of the 1974 Costs of Indian Point and Alternatives," August 1975.